Praise for *The Glutathione Revolution*

"How exciting it is that my distinguished colleague Dr. Nayan Patel has produced this book on one of the most fascinating nutrients—glutathione! In this valuable book, he clearly presents solid, up-to-date scientific information about glutathione and its benefits for health and anti-aging. We have long needed such a book. I am confident that, in reading it, you will learn much about glutathione in general and will discover many practical clues for optimizing your glutathione levels. Greater well-being and increased longevity will be the welcome result."

— DOMINIQUE M. FRADIN-READ, MD, MPH,
founder of VitaLife-MD Integrative Medical Practice

"I am so pleased Dr. Patel has put together a cohesive book on the current research and benefits of glutathione! Thirty years ago I started as an ER physician, then specialized in laser medicine, then evolved into Integrative Medicine, and now specialize in Regenerative Medicine. At each stage of my development, the importance of glutathione as a fundamental component of health and well-being has been reinforced. We offer and utilize glutathione IV pushes and other IV nutrition solutions at my AMA Regenerative offices in Orange County and Beverly Hills, California, and I am thrilled to offer more glutathione resources to my patients!"

— ALICE PIEN, MD, Medical Director, AMA Regenerative Medicine

"Being a sportsman and a health-conscious individual, I am always on the lookout for something to better my health. Thanks to Dr. Patel's book, the benefits of glutathione are now available to the world."

— NIKHIL CHOPRA, international cricketer and commentator,
India 1999 Cricket World Cup team

"Pharmacists help people live better, healthier, safer lives. Pharmacist Dr. Nayan Patel's book foretells a revolution in focusing on healthy behaviors for maximizing your stores of this invaluable personal biological asset. The book is a gift of knowledge, encouragement, and hope to humanity. Here's to better health!"

—RONALD P. JORDAN, Dean of School
of Pharmacy, Chapman University

"This book is a complete game changer when it comes to people's health. Having the lifestyle I do—with nonstop filming schedules, business meetings, and endless traveling—I am so, so glad that I was exposed to the benefits of glutathione a few years ago. But in a world where information comes hard and fast, it is very difficult to decipher what works and what doesn't, what to take, and what not to, when it comes to your health. This book by Dr. Patel makes it a lot easier to decide—and the best thing to start with is glutathione. The secret is out, thanks to Dr. Patel and the wonderful book he has written!"

—SUNIEL SHETTY, actor, health and fitness pioneer,
businessman/entrepreneur, Mumbai, India

"In the era of unheralded biological hazards and everyone adjusting to the 'New Normal' with the COVID–19 pandemic and with daily tenseness, *The Glutathione Revolution* is a triumph, with the aim to 'add life to years and not just years to life.' This deep dive into the science of glutathione will start a rave in the world of antioxidants. With simple and easy-to-follow action plans for detoxing, *The Glutathione Revolution* unleashes the secret of being healthier, looking younger, and adopting a healthy lifestyle."

—SANDEEP GUPTA, director and CEO, Nutraworks, chief founder
and director, Expert Nutraceutical Advocacy Council

"The two most metabolically active organs in human physiology are the liver and brain—and it so happens that these two organs generate and utilize more glutathione than any other organs. The challenge has always been how to supplement our bodies with glutathione—even the best and purest sources are not easily absorbed orally, necessitating a visit to your doctor for IV treatment. Now, everyone can have optimal glutathione levels every day!"

—ASHER MILGROM, PhD, CSO, AMA Regenerative Medicine

"In the fast-paced world of rapidly evolving technology, from quantum computing to space exploration, the human body and its systems still remain mysterious. Yet understanding our bodies means harnessing our greatest tool to help us become the best versions of ourselves. As a believer in continuing our education and evolution as humans, I was greatly encouraged and excited to learn from Dr. Patel's much-needed comprehensive book on glutathione, supporting the quest to learn everything we can about this unrivaled and life-giving superantioxidant. Cheers to a bright future full of space travel, better lives on Earth, and plenty of glutathione!"

—Eric Anderson, CEO, Planetary Holdings, cofounder and chairman, Space Adventures

"The Glutathione Revolution is the absolute need of the hour, as we are seeing an increasing number of patients around the world suffering from ailments that can be prevented or cured with the consumption of antioxidants like glutathione."

—Dr. Milind A. Antani, MS, LLB, leader, Pharma, Healthcare, Medical Devices, and Nanotechnology Law Practice, Nishith Desai Associates

THE
GLUTATHIONE
REVOLUTION

THE GLUTATHIONE REVOLUTION

Fight Disease, Slow Aging,
and Increase Energy
with the Master Antioxidant

NAYAN PATEL, PHARM.D.

hachette
BOOKS

NEW YORK

Copyright © 2020 by Nayan Patel
Jacket design by Terri Sirma
Jacket illustration © StudioMolekuul/Shutterstock.com
Jacket copyright © 2020 by Hachette Book Group, Inc.

Hachette Go, an imprint of Hachette Books
Hachette Book Group
1290 Avenue of the Americas
New York, NY 10104
HachetteGo.com
Facebook.com/HachetteGo
Instagram.com/HachetteGo

First Edition: September 2020
Hachette Books is a division of Hachette Book Group, Inc.
The Hachette Go and Hachette Books name and logos are trademarks of Hachette Book Group, Inc.
The publisher is not responsible for websites (or their content) that are not owned by the publisher.

Print book interior design by Linda Mark

Library of Congress Cataloging-in-Publication Data has been applied for.

ISBNs: 978-0-306-87397-3 (hardcover); 978-0-306-87395-9 (e-book)

Printed in the United States of America

LSC-C

10 9 8 7 6 5 4 3 2 1

This book is dedicated to those who desire
to be their own health advocate, are willing to
ask questions, and seek the answers for themselves.

CONTENTS

FOREWORD

WE HAVE LONG KNOWN THAT GLUTATHIONE (GSH) PLAYS AN INTE-
gral role in detoxing our systems and helping to prevent disease.
But by *we* I mean the medical establishment; unfortunately, very few non-
medical people have ever heard of glutathione. So I'm grateful that in this
book pharmacist and researcher Nayan Patel is not only bringing glutathione to wider attention, he is also helping us understand how it works, where
the GSH research has been—and where it's heading—and, most important,
how we can get more of it.

We live in an era of unprecedented biological hazards (the rampant
COVID-19 virus) as well as daily environmental stressors—some of us are
being bombarded by more toxins than our bodies can fight off. For that
reason, this book is not only timely, but crucial. And all the better for being written by someone who has done important work in stabilizing the
actual glutathione molecule and who is also on the GSH front lines, con-
sulting with patients and doctors and answering questions about the antioxidant on a daily basis.

I was an early advocate of glutathione. For over thirty years now, I've
been talking about the importance of this little-known molecule and doing my best to let people in on the news. In 2010, for instance, I wrote in
the *Huffington Post* that glutathione is "the secret to prevent aging, cancer,

heart disease, dementia, and more, and necessary to treat everything from autism to Alzheimer's disease." That's an assessment I still stand by today, but as Dr. Patel demonstrates in his clear-eyed, evidence-based paean to GSH, that's just the tip of the iceberg. There is a lot more to discover about GSH, and it's all here on the upcoming pages.

As a practitioner of functional medicine, I believe in empowering people to take charge of their own health. So I was happy to learn that Dr. Patel would not only be helping readers understand the many ways that glutathione affects our well-being, but also informing us how we can maximize GSH in our bodies. The brilliant thing about glutathione is that we make it ourselves—how great is that? And yet the onslaught of factors I described earlier, plus just the mere fact of aging, can cause a glutathione lag. We need to know how to head off that lag and increase our GSH arsenal, and Dr. Patel tells us how.

Many years ago, as I tackled my own health challenges, I experienced glutathione's powers firsthand. I am therefore thrilled that the conversation about glutathione is continuing, now in depth, and led by the steady hand of Dr. Patel.

Welcome to the Glutathione Revolution.

Mark Hyman, MD

INTRODUCTION

WHAT IF I TOLD YOU THAT THERE'S A POWERHOUSE SUBSTANCE THAT has the capability to both protect your body against disease and help it heal if you do get sick?

What if I also told you that this same health-enhancing substance can help slow down the aging process, refresh your energy level, and keep your skin looking luminous?

What if I told you, too, that this exceptional substance is far better at detoxing your body than any $90 cleanse kit on the market?

And here's the kicker: What if I told you that this incomparable substance is one of the most abundant molecules in your body, second only to water—and that you make it yourself?

Have I got your attention yet?

Glutathione is the disease-fighting, age-slowing, energy-enhancing, detoxing, and beautifying antioxidant that you've never heard of . . . or perhaps you have. Each day, the number of individuals who are getting wise to glutathione's benefits grows. People, search metrics show, have been Googling it like crazy! Even so, aside from scholarly articles (and there are thousands of them—glutathione is very well studied), there's little accessible yet authoritative information out there explaining how integral this mighty molecule is to nearly every physiological function. And, just as important, few people know about the best ways—including all the natural ways—to increase glutathione levels in the body. With this book, I hope to change all that—and I think you're going to be amazed by everything you learn about this potent antioxidant.

GSH—shorthand for *glutathione*—is produced in every cell in the body and it's our first line of defense against, well, just about everything. Among glutathione's most important jobs is shielding us from the day-to-day assault of free radicals, rogue molecules that do harm to all facets of the cells, setting the stage for disease and accelerating aging. GSH also grabs on to waste products and environmental toxins, such as pesticides and metals that surreptitiously enter our system, helping to flush them out of the body.

But that isn't all. What makes glutathione such a superantioxidant—and why it's interchangeably called "the master antioxidant" and "the mother of all antioxidants"—is its ability to rebuild both itself and other essential antioxidants, such as vitamins C and E. There's no other substance in the body quite as agile and far-reaching as glutathione, and its biochemical heroics allow it to play a leading role in everything from protecting against heart disease, diabetes, cancer, and Alzheimer's (among other conditions) to keeping skin clear and bright.

Except for the rare few, human beings (animals and plants, too) are born with plenty of glutathione. However, most people's ability to produce glutathione declines after about the age of twenty and drops with each passing year. By the time you reach your forties, your GSH levels may drop by as much as 20 percent. That may not seem like a considerable handicap, but it can be if your body's demand for glutathione is high. And it very well may be: Environmental insults, such as air pollution, toxins lurking in products you use, junk food eating, smoking, drinking, overwork, overworrying, and even excessive vigorous exercise, can send your need for glutathione through the roof. So, the less you have, the less likely you'll be to adequately deal with threats to your well-being. Over time, and depending on how low it goes, lack of glutathione can make the body more vulnerable to a long list of woes.

And yet, it doesn't have to be that way. There's good evidence to suggest that you can maintain ample glutathione in your body by adopting lifestyle habits that both preserve and build the antioxidant. This can lead to all manner of improvements in how you look, feel, and even how long you live. How long? Consider a study conducted by Dutch researchers.

There are many theories about why some people are able to live into their centennial years. For this particular investigation, the scientists set out to see whether it may have something to do with having a superior antioxidant defense system, and more specifically a superior ability to produce the enzyme that keeps glutathione levels high. The study involved forty-one men and women aged 100 to 105 years, whose blood levels of glutathione reductase (the GSH-catalyzing enzyme) were compared to those of fifty-two "kids" (people aged 60 to 79 years). Despite the age difference, the centenarians had unexpectedly higher levels of glutathione reductase than did the younger men and women. The Dutch researchers concluded that the enzyme, a critical part of the body's glutathione system, seemed related to the centenarians' long-term survival.

To the scientific community, the relationship between glutathione and longevity isn't news (the Dutch study, in fact, was done in 1998). But the rest of the world has little idea about what researchers have known for a long time: Glutathione can be life changing. I believe, though, that as word gets out, it will start a new wave of thinking about maintaining good health and fighting aging. I didn't call this book *The Glutathione Revolution* for nothing!

———

IF GLUTATHIONE PLAYS SUCH AN OUTSIZE ROLE IN OUR WELL-BEING, WHY, you might ask, hasn't it enjoyed the spotlight like other important antioxidants (vitamins C and E, for example)? The main reason is because the glutathione molecule's size and instability have made it difficult to develop effective supplemental forms. So, while there have been GSH supplements on the market for some time, their benefits are questionable. Not surprisingly, health-care professionals have been hesitant to recommend something that, although harmless, would not likely offer many advantages.

Fortunately, that's changing now, something I can say with confidence since I, along with my research partner Dr. Chinh Tran, worked to shrink and stabilize the glutathione molecule, leading to the development of a form of GSH that's easily absorbed through the skin. We patented it and have produced an over-the-counter glutathione solution (along with other topical products) that are now used by thousands of people to address various health and cosmetic issues. These days, it's also easier than ever to find intravenous (IV) glutathione treatments at spas, clinics, and doctors' offices. When administered by a reputable source (preferably in a medical setting), these drips are safe and effective for short-term therapy and are something that famous faces and people of means have been using for years. But with the increase in availability (and moderation of price), you don't have to have a celebrity's cash reserves or insider info to find and afford an IV glutathione treatment anymore.

The other reason I believe that glutathione has stayed under the radar for so many years is that there's been little written about how to *naturally* increase your GSH levels. Everything from the foods you choose to the sun protection you use can make a difference in your glutathione levels. In this book, I'll map out easy lifestyle strategies to help you kick your glutathione-making machinery into gear. You'll learn what to eat—and what not to eat—as well as how such things as exercise and a short meditation can give you a GSH boost. Depending on what you're trying to achieve, there may be times when you'll need to add supplementary forms of glutathione (or supplementary forms of glutathione building blocks) to your regimen. I'll give you strategies for this approach, too, so that you can make the choices that will help you meet your goals.

———

I AM A FATHER, A HUSBAND, A SON, A BROTHER, A PHARMACIST, AN EDUCA-tor, and a researcher. My roles in this life are many, but I believe my mission on this earth is to be a facilitator of wellness and to provide ways for society to learn about how the body can heal itself. Therefore, I am most looking forward to working with international medical and philanthropic communities to bring opportunities for healing and good health to every corner of the globe. It's a passion that you might say is part of my inheritance from my father.

As a young married man, my father moved from India to Zambia to start a farm. The local community was severely lacking in food and my father's goal was that his farm would help create a food supply that would eventually be self-sustaining. My dad never had a lot of money, just enough to survive. Even so, he was able to build a thriving farm that served the community on about 2,500 acres of land.

My father worked many years on this farm, but once it was fully built out with infrastructure, crops, and machinery, he walked away from it and gave it to the community. Yes, he *gave* it away. He felt like it always belonged to the people, so he gave it to them. And because of my father's approach to life, he is a very happy, fulfilled, and peaceful man. Some might call him simple; I just call him wise.

My dad made me promise to never lose sight of the fact that my life has always been and will always be about and for the people. Recently, he reminded me that my job on this planet was to be a healer and that our family is here to give back. Accordingly, I consider my mission statement in life to be "Be educated and educate others." By passing on what I have learned about glutathione to you, I hope to fulfill that mission.

If you are seeking a way to restore balance to any areas of your life that lack harmony, I believe that you will find those answers in glutathione. One of my favorite sayings is, "What you are seeking is seeking you." GSH has been looking for you—and now you are found!

GLUTATHIONE IN A NUTSHELL

THE STORY OF GLUTATHIONE is that every cell in your body benefits from it and when your glutathione pools go down, the damage to your cells goes up. It's straightforward: Glutathione works against a variety of issues that people encounter either through self-induced damage like lack of taking care of themselves, not getting the right nutrients, or through environmental burden.

—JAMES B. LAVALLE, RPh, ND, clinical pharmacist, certified nutritionist, author of *The Metabolic Code*, and integrative medicine program director for the NFL Hall of Fame Health and Performance Program

GLUTATHIONE AT A GLANCE

EVERYWHERE I GO, I get peppered with questions about glutathione, not only because most people still don't know much about it, but because they're also hungry for news that can help them feel and look their best. You're going to learn a lot about glutathione in the pages of the book, but here are some quick answers to whet your appetite for more information.

What Is Glutathione?

Glutathione is an antioxidant, a detoxifier, and the second most abundant molecule in the body. It's your first line of defense in fighting free radicals, makes it possible for your body to eliminate toxins, and supports overall cellular health. Oh, and by doing all this, it also helps defend against disease and slow down aging.

Where Does Glutathione Come From?

You make it! That's the beauty of glutathione; it's natural. Your body combines three amino acids to create it. It also gets some from glutathione-containing foods (see page 142).

If I Make It, Why Are There Supplementary Forms Available?

For two reasons. The first is that, as with most biological processes, the production of glutathione slows down as the body ages, beginning in your twenties. However, if you do certain things—such as eat right, exercise moderately, and avoid as many toxins as possible—you will not likely need supplementary glutathione at all. That said, and reason number two, in this modern world, the body of most people is subjected to a lot of environmental stressors. Even the shampoo you use may be a source of chemicals that your body will have to work extra hard to get rid of. And it's this extra work that causes the problem: The greater the assault on your body, the greater the chance that you will max out your glutathione stores.

How Do I Know If I'm Deficient in Glutathione?

If you're having symptoms, anything from fatigue to stomach problems to weight gain to dull skin, you can have your physician order a glutathione test to assess your level. Many people, though, simply try ways to increase their glutathione levels either through natural or supplementary means (or both) to see whether their symptoms or their energy and vitality improve. It's like cleaning your glasses when your eyesight is fuzzy. There's no risk involved—you can't overdose on glutathione—and there's a good likelihood that it will clear up the problem.

What Kind of Supplementary Forms of Glutathione Are Available?

There are oral forms of glutathione, but they are not well absorbed. IV glutathione is considered the most effective way to supplement, but it's expensive and time consuming if you have to do it on an ongoing basis. The best and easiest way to supplement glutathione is through topical forms.

What Will I Improve by Raising My Glutathione Levels?

Just about everything. If that sounds like an exaggeration, consider that almost every major disease, and even some minor ones, is related to oxidative stress—an imbalance of damaging free radicals and the antioxidants that demolish them. Consider, too, that your liver cannot do its best work without adequate glutathione. Glutathione is involved in just about every function in your body, from your immune system activities to your cells' ability to churn out energy. This is one case when more is definitely more!

........•........

PART I

Meet the Mother
of All Antioxidants

1 — WHAT IS GLUTATHIONE?

I'M A PHARMACIST, SO I HAVE GREAT APPRECIATION FOR ALL THE THINGS we can take to improve our health and even save our lives. It's hard to picture modern civilization without the prescription drugs that have made it possible for us to live considerably longer than our ancestors. Or the over-the-counter medicines that cure and comfort, not to mention the vitamins, minerals, and herbs that boost vitality and offer protection against illness. These remedies are some of the most ingenious and important discoveries human beings have ever made.

But as much as I value all the elixirs, pills, and potions produced in a laboratory, I am even *more* in awe of the remedies our body manufactures by itself. In most cases, we are born with the ability to make disease-fighting, detoxifying agents that so elegantly and effectively protect us from harm that we don't even know they're working! Every minute of every day, these self-generated properties endeavor to ensure that we avoid all kinds of illnesses, be it of the merely bothersome sort (say, the common cold) or something very serious (cancer being a prime example). *Our own body makes them!* Incredible.

And what is first among these physiological wonders? An antioxidant known as glutathione.

Glutathione, or if you want to get technical about it, glutathione sulf-hydryl, GSH for short, is found in every cell in the body. It's ubiquitous because it's so important. As one of nature's most powerful antioxidants—and I'll explain just exactly what antioxidants do further on in this chapter—glutathione is extremely versatile, taking on many jobs. It is also first on the front lines of our defense against molecular marauders that damage DNA and other cellular matter, helping to prevent disease and accelerated aging. Glutathione is a detoxifier too. There's an especially large concentration of the compound hanging around in the liver, where it assists the body in eliminating waste and potentially poisonous substances. And as a player in the immune system, GSH enhances the production and activity of the cells that wipe out bacteria, viruses, and other invaders.

To my mind, it's truly a wonder that glutathione is not better known. It's the body's workhorse, toiling away, getting little of the glory heaped on other antioxidants, such as beta-carotene and vitamins C and E (though, of course, they're important too). But that, thankfully, is changing. I think that as you learn more about the mechanics of glutathione, you'll understand not only how underappreciated it's been, but how worth your while it is to take steps to heighten your GSH levels.

HOW THE BODY BUILDS GLUTATHIONE

Glutathione is new to a lot of people, but researchers have known about it for a long time. It was first discovered in 1921 by Sir Frederick Gowland Hopkins, a former insurance clerk who segued into science and ultimately became chair of the biochemistry department at Cambridge University. Hopkins isolated GSH in yeast (the compound exists in plant as well as animal cells), setting the stage for other scientists to uncover the many roles it plays in keeping the body up and running. Hopkins, by the way, was also known for isolating tryptophan, an essential nutrient we derive from food, and he was awarded a Nobel Prize in 1929 for his work establishing the important role vitamins play in physiology.

So, what exactly was it that Hopkins saw under his microscope all those years ago? A very simple protein. Proteins play all kinds of different roles in the body. You probably already know that they are integral to creating muscle and hair and the collagen that gives our skin structure. But some proteins function as antibodies, some catalyze chemical reactions, and some act as messengers. And some, like glutathione, are antioxidants.

Amino acids, as you may remember from high school biology, are the building blocks of all proteins. To create glutathione, the body must string together three particular amino acids: glutamic acid, cysteine, and glycine. There is no substitute for any of these ingredients. All proteins have a biological blueprint that must be followed to the letter or they will not be able to do their jobs. In other words, it's not like a recipe for spaghetti marinara, where you can leave out the red pepper flakes, or add the salt at the end instead of the beginning of cooking, and still have a perfectly fine dish. To build glutathione, you must have all of its components on hand and assemble them in the proper sequence.

Fortunately, glutathione needs only three amino acids, unlike, say, human growth hormone, which requires a whopping 191. What's more, the body can make glutamic acid, cysteine, and glycine itself. And yet, as we age and the tiny manufacturing plants in our body slow down, and during times when glutathione is in high demand (needed, say, to help purge the remnants of too many gin and tonics after a birthday celebration), we can't synthesize enough glutamic acid, cysteine, and glycine to fill the order. That's when dietary sources of the three amino acids become particularly important. In subsequent chapters, I'll talk more about how food affects your glutathione levels, and give you tips on how to adjust your diet to maximize GSH production.

IS IT A BIRD, A PLANE . . .
OR A SUPERANTIOXIDANT?

A LITTLE ASIDE HERE about semantics.

I am often asked whether glutathione is an antioxidant or a protein or a tripeptide or an enzyme or a molecule or a compound. The answer is yes, yes, yes, yes, yes, and yes.

Why so many names? Glutathione has many different jobs in the body—the reason it's so essential!—and structurally it falls into many different categories. Here's a little primer that will help put things into perspective.

Antioxidant—A substance that neutralizes free radicals, preventing oxidative damage. Glutathione's number one job is to lend electrons to charged oxygen molecules so they don't steal them from other important structures, such as DNA.

Protein—Large molecules consisting of a chain of amino acids. Protein makes up most of cell matter and is involved in everything from muscle contractions and immunological responses to enzymatic and hormonal actions. Because it's made up of three amino acids, glutathione is a protein.

Tripeptide—A peptide is a particular type of amino acid chain formed through the elimination of a water molecule. Glutathione is a tripeptide because it's made up of three amino acids: glutamic acid, cysteine, and glycine.

Molecule—A group of atoms formed by chemical bonds.

Compound—A substance made up of two or more materials. In glutathione's case, it's a substance made up of glutamic acid, cysteine, and glycine.

Despite all the names that glutathione can go by, in this book I'll be primarily referring to it as an antioxidant. That's because I think about compounds the same way I think about people: it doesn't matter who you *are*; it matters what you *do*. So, while it's interesting to know what chemical category glutathione falls into (especially if you love chemistry, like me), we're all more interested in how GSH functions in the body, right? Not forgetting that glutathione is also an essential detoxifier, let's settle on a name that describes its greatest superpower: neutralizing free radicals to prevent oxidative stress. From here on out, you'll mostly see me refer to GSH as an *antioxidant*.

........●........

MEET THE ENEMY: FREE RADICALS

Not too long ago, nobody but medical professionals and avid readers of health literature had even heard of antioxidants. But, these days, judging from food, supplement, and beauty-product advertisements and labels, most people not only know what antioxidants are, they're actively seeking them out. It's not unusual to see a manufacturer boasting of its corn chips' "antioxidant power" or the "antioxidant advantage" of its orange juice. Beauty products are regularly touted as "antioxidant treatments," and antioxidant supplements fill the shelves of mainstream pharmacies and natural grocery stores alike. That's a good indication that most of us know that antioxidants like glutathione are good for us.

So, what do they do?

Biochemists have defined *antioxidant* as "any substance that delays, prevents, or removes oxidative damage to a target molecule." They do this by neutralizing and disarming harmful molecules called free radicals. This is both a full-time job—free radicals are constantly on the warpath—and an essential one. Perhaps the antioxidant expert Lester Packer said it best: "As the adage goes, wherever there is smoke, there is fire. Similarly, wherever there is disease and destruction, there are free radicals."

The main (though not the only) cause of free radicals in our body is . . . our body. Or more specifically, chemical reactions involving oxygen that take place in our body. One place that free radicals are typically generated is in the mitochondria, tiny structures inside cells where oxygen molecules are converted into energy. During the conversion process, some oxygen molecules end up losing one or more of their electrons, leaving them unstable, reactive, and a threat to other molecules in their path.

A regular, harmless oxygen molecule is encircled with an equal number of negatively charged particles (electrons) and positively charged particles (protons). This balanced pairing renders the molecule neutral; it's neither positively nor negatively charged, and happy just the way it is. But when the tumult of a chemical process causes one or more of its electrons to go missing, the molecule—now a free radical—gets as purposeful as a member of the lonely hearts club swiping right on Tinder: It's highly motivated to find a soulmate for that unpaired proton, and it wants to do it fast.

What follows are acts of theft. Because the imbalance of protons and electrons has left the free radical electrically charged, it has the ability to easily steal electrons from other molecules, and steal it does—a process called oxidation. But it doesn't end there. The robbed molecules then cannibalize the electrons of their neighbors, who then snatch away the electrons of *their* neighbors. The chain of theft can go on and on, leaving damage in its wake. If you want to have a picture in your mind of what oxidation is like, think of rust. In the presence of moisture—H_2O—iron loses electrons to oxygen molecules, creating the corroding we have come to know as rust.

A number of things besides energy production increase the formation of free radicals in the body. Exercise, for all its benefits, ramps up the creation of the errant molecules as a by-product of fat burning. (This isn't necessarily all bad; see Chapter 6). When the immune system snaps into action it, too, through various pathways, elevates free radicals' presence in the body. Such things as sun exposure, environmental pollutants (smog, cigarette smoke), food additives, pharmaceuticals, and pesticides can also produce an upswing in their formation. Worse, many of these invaders contain free radicals themselves, magnifying the problem as they dump them into your system.

It should be noted that free radicals aren't completely villainous. They are integral to helping the immune system fight viruses and bacteria and are even generated by certain chemotherapy drugs as part of the plan to treat cancer. One study showed that they may help heal wounds too. Plus, our body provides a counterpart to free radicals: antioxidants, which are nature's way of making sure the harmful effects of free radicals can be modulated. But when free radicals outpace antioxidants—not an uncommon occurrence given the realities of modern life—the imbalance creates something called oxidative stress. Oxidative stress is a result of having more free radicals than your body can get rid of, and that's where the trouble begins.

A long list of diseases are associated with oxidative stress and the chronic inflammation it can cause. Cancer is one of them. Free-ranging free radicals can damage DNA, causing cells to mutate and become cancerous. (Didn't I just say that free radicals can help cure cancer? Yes, but oddly enough, they can both cause and cure tumor development, depending on what molecules they're targeting. When the target of free radicals is healthy DNA, they're harmful; when the target is a cancer cell, they're helpful.)

Free radicals also have the ability to oxidize fats in the body, leading to deposits on artery walls and making the arteries harden, both precursors to heart disease. There's an oxidative stress link to just about every malady you can think of, including stroke, Alzheimer's disease, diabetes, rheumatoid arthritis, and macular degeneration. Free radicals are known to accelerate aging too. The lines on your face as you grow older may be due to oxidative stress helping to spur the breakdown of the proteins that give the skin its plumped-up shape and keep it looking smooth. Free radicals can't get all the blame for wrinkles, but there's demonstrable evidence that they speed up the wrinkling process.

There is also evidence that oxidative stress shortens telomeres, another way free radicals contribute to aging. Telomeres are most often described as caps composed of protein and DNA that sit on the ends of chromosomes (think of the little plastic thingy on the ends of a shoelace and you'll have a pretty good picture of a telomere). As the years go by and cells divide, telomeres become smaller. When they get small enough, they tell their cells to stop dividing, effectively hampering the body's ability to regenerate and refresh itself. Hastening this natural process is another way free radical damage makes us not only more vulnerable to disease, but more likely to suffer the deterioration of our tissues that comes with aging. I'll tell you more about telomeres in Chapter 5.

ANTIOXIDANTS STRIKE BACK

As I said before, the body doesn't just let free radicals run rampant without fighting back. Antioxidants regularly neutralize free radicals, stopping them in their otherwise injurious tracks. And as long as they're in ample supply, antioxidants retain the edge over free radicals.

There are hundreds of different types of antioxidants. You have probably heard of vitamins C and E, beta-carotene, and, maybe, coenzyme Q10, which all have the power to squelch free radicals and prevent oxidative stress. Maybe you're even aware of some of the less well-known antioxidants, such as selenium and alpha lipoic acid. (Also see "Glutathione's Brothers in Arms," page 14.) Some of these compounds, vitamins C and E among them, can only be obtained from food; your body doesn't make them. Other antioxidants, such as glutathione, the body can synthesize itself.

Antioxidants attack free radicals in a few different ways. One is by preventing their formation in the first place. Glutathione in its role as a detoxifier does this by removing many toxic elements from the body before they can create free radicals. It's the body's ounce of prevention. The other way antioxidants protect against damage is by donating their own electrons to free radicals so that the charged molecules don't steal them from DNA and other essential compounds. Antioxidants have electrons that they *want* to give away. Once they do and the free radicals are turned into harmless water molecules, those reactive atoms cease their pilfering and pillaging, and damage is averted.

While all antioxidants are generous in this way, glutathione is by far the most magnanimous. Say you and I are sitting in a room with a few other people and someone comes in and says, "Can anybody spare a hundred dollars?" You're the first one to pull out your wallet, and even when others offer to contribute, you say, "No, no, I got this!" That's how glutathione behaves in the presence of other antioxidants. If it's sitting next to a selenium molecule, the selenium isn't going to lose a single electron, because GSH is on the job.

Still, one antioxidant can't do it all. Going back to the hypothetical room where you and I sit, say that once you have loaned out the $100, a second, then a third, then a fourth person comes along and makes the same request. That's when everyone else in the room has to pull out their wallets too. And so it is with the band of free radical scavengers. Glutathione, as effective as it is, can't be the end-all, be-all, so other antioxidants must contribute as well. A lot of who does what depends on how much GSH there is in the body at the time and where the interactions are taking place. Different antioxidants are concentrated in different parts of the body, and that can affect which of the defenders end up donating their electrons.

The way antioxidants work in the body is a little complicated. It's not like a game of Mousetrap, where the red paddle hits the bucket that sends a ball rolling down an alley, which causes a bigger ball to glide through a bathtub, falling on a diving board and launching a swimmer into a pool, which then activates the mousetrap—over and over again. Success at Mousetrap, in other words, always involves the same cascade of events. With antioxidants, the chain of action is not always the same. Neutralizing free radicals can work in several different ways.

We need antioxidants, such as vitamins C and E, on the front lines of the body's defense system. So, after those soldiers have given up their electrons to stabilize free radicals, they don't just end up spent, lying in the body's trash heap. Nor do they become destructive free radicals themselves (which would, of course, defeat the whole purpose of their work). Instead, they go through a process called recycling, and glutathione is integral in this endeavor.

When another antioxidant, say, vitamin C, hands off an electron to placate a roving free radical, glutathione replenishes the C molecule with one of its own electrons. Now the C molecule is good to go back to work, but the glutathione molecule is missing an electron. So what does it do? It replenishes itself by fusing with another electron-deficient GSH molecule. Together they create a new, stable glutathione molecule. In essence, glutathione not only goes to war against invaders, it reloads itself, rearms its fellow soldiers, and prepares to meet the next threat.

But sometimes the role is reversed, and my personal theory is that it's reversed much of the time. As the most important antioxidant—it's known as *the mother of all antioxidants* (see "The Mother or the Master of All Antioxidants? Both" on page 13)—glutathione needs to stay in the game. We need as much of it as possible. But in the metabolic sense, glutathione is very expensive to make. It "costs" the body enzymes and energy and other elements to make. There is, however, a cheaper way to keep the coffers full of GSH and that is to recycle it. The body can do that all day long. And how does glutathione get recycled? It does it itself by grabbing electrons from other antioxidants, such as vitamins C and E. These antioxidants are themselves low cost. The body doesn't have to make them; it gets them from food. So, even though these other antioxidants do work of their own, my theory is that their main role is to help glutathione recycle itself. Here is how it works:

A positively charged, reactive oxygen molecule
(free radical) comes into view

A glutathione (GSH) molecule tosses a negatively
charged hydrogen electron its way

The free radical gratefully accepts the hydrogen
electron and transforms into a harmless,
neutral water molecule (H_2O)

The GSH molecule, now positively
charged itself, seeks out another positively
charged GSH molecule

They merge, creating a neutral form
of glutathione called GSSG

The GSSG molecule lies dormant until
it's time to go to work again: GSSG splits
and grabs electrons from lesser antioxidants,
such as vitamins C and E

Two recycled GSH molecules begin the
battle against free radicals all over again

Glutathione can't always be recycled—something different happens when GSH is functioning as a detoxifier, which I'll get into in Chapter 4— but a great deal of it can and that helps keep the wheels of the free radical defense system turning.

THE MOTHER OR THE MASTER
OF ALL ANTIOXIDANTS? BOTH.

MANY SCIENTISTS REFER TO glutathione as *the mother of all antioxidants* or, sometimes, *the master of all antioxidants*. And glutathione does quite a bit of what we might call "mothering": It makes sure that the little ones—like vitamin C, beta-carotene, and other antioxidants—are taken care of after they have donated their electrons to neutralize free radicals. When GSH replaces those donations with its own electrons, it's like the selfless mom who, when the temperature drops, gives up her sweater to the ten-year-old who left hers at home.

In a different sense, glutathione acts in a masterly fashion. As the only antioxidant that can recycle itself as well as the antioxidant that decides when to sacrifice its brethren by snatching their electrons, it runs the show. And that's essential, because no other antioxidant can go all the places in the body that glutathione can go. We need it more than any other free radical slayer. We simply can't do without it.

BEYOND ANTIOXIDANT ACTIVITY

In subsequent chapters, I will tell you more about glutathione's role in keeping your body healthy. But here is a little preview. Aside from its antioxidant activity, one of GSH's main jobs is to help detoxify the body. Glutathione is concentrated in the liver, the organ that's a clearinghouse for toxic substances. The body is continually confronted with harmful compounds, some that are natural by-products of things you might not think of as unsafe (e.g., medications) and some that are unnatural by-products of modern life (e.g., ozone and fine particles in smog). When glutathione meets up with toxins, it deactivates them in one of two ways: either by neutralizing them and turning them into water molecules, or by helping to eliminate them from the body.

GLUTATHIONE'S BROTHERS IN ARMS: FIVE OTHER IMPORTANT ANTIOXIDANTS

Just about everything that happens in the body is a product of complex chemical reactions involving many players. In other words, there's a lot of teamwork going on. Glutathione gets considerable help from—and offers a lot of assistance to—other antioxidants. In the following chart, I outline GSH's most important teammates, along with what foods will help ensure that you get enough of them.

ANTIOXIDANT	ROLE IN THE BODY	WHERE IT COMES FROM
Vitamin C	A free radical scavenger that helps control infections; is used to build skin, blood vessels, and other parts of the body. Also helps the body absorb iron.	The body doesn't make or store vitamin C. You can get it from citrus, strawberries, cantaloupe, tomatoes, broccoli, and other cruciferous vegetables.
Vitamin E	A free radical scavenger that helps bolster the immune system and is integral in the formation of red blood cells.	The body doesn't make vitamin E. It's available from vegetable oils, nuts and seeds, and green leafy vegetables.
Vitamin A (retinol, beta-carotene)	Helps with the production and activity of white blood cells; participates in bone remodeling; regulates cell growth and division.	Different forms of vitamin A are derived from different sources: some from animal products and fortified foods, others from plant compounds in leafy greens and orange or red produce.
Lipoic Acid (also known as alpha lipoic acid)	Free radical scavenger; recycles other antioxidants and helps restore glutathione when lacking; turns off genes that accelerate aging and cause cancer.	Lipoic acid occurs naturally in the body. Small amounts are also available in potatoes, spinach, and red meat.
Coenzyme Q10	A free radical scavenger that also recycles vitamin E and is essential in the process of creating energy.	Coenzyme Q10 occurs naturally in the body. Small amounts are also available in fish (e.g., tuna, salmon, mackerel, and sardines), vegetable oils, and meats.

Our immune system also relies on glutathione. In fact, macrophages—white blood cells that scavenge infectious invaders—don't leave home without GSH. Think of it this way. If you were a firefighter called to fly up to Alaska and fight a big old raging blaze, you would make sure to bring your tool bag. And you would make sure that the tool bag had all the flame-battling gear you needed, including, say, your ax. Glutathione is the macrophage's ax, an instrument in its tool bag that's going to help it do its job properly.

Now you know all that GSH does to keep your body healthy. But what happens when your glutathione levels drop? I'll tell you all about it in Chapter 2.

2 — WHEN GLUTATHIONE GOES MISSING

WHEN NEWLY ARRIVED IN THE WORLD, MOST CHILDREN COME EQUIPPED with everything they need to thrive, and that includes an army of glutathione and other antioxidants in their cells. If they're lucky, they also have parents looking out for their health—bundling them up to keep out the cold, slathering them with sunscreen to fend off UV rays, and feeding them nutritious foods to promote growth.

Still, even if they have the most conscientious parents, kids can develop—and get away with—some seemingly detrimental habits. They may eat lots of sugar and be so finicky they're practically living on macaroni and cheese from a box. They may spend hours playing outdoors in the heat and sun. Other than a few dental cavities, the immediate ramifications of these habits can be relatively small. Teenagers can even drink alcohol (illicit though it may be) without feeling much of a hangover. Many of us look back with wistful longing for those days when we could eat and drink just about anything we wanted, and our skin still seemed fresh and dewy even if we had been out in the sun. Ah, youth! Ah, a fully stocked arsenal of glutathione!

When you become an adult, there are lots of good reasons to turn over a new leaf and set aside any bad habits you once got away with. And one of the most significant reasons is that glutathione's protective power diminishes as you grow older, making you more susceptible to disease and accelerating both visible (such as lines and wrinkles) and invisible (such as a lesser ability to withstand heat and cold) signs of aging. Two things contribute to this change. One is that the natural production of glutathione slows down; you simply don't make as much GSH. The other is that your need for the mother of all antioxidants may go up—especially if you never gave up the bad habits you developed when you were younger (or if you adopted new transgressions as an adult). But even if your lifestyle is fairly healthy, there is still a world full of free radicals and toxins out there waiting to do damage. Speaking to the magazine *Experience Life*, David Perlmutter, MD, author of the books *Power Up Your Brain: The Neuroscience of Enlightenment* and *Grain Brain*, put it perfectly: "We are in a situation where our manufacturing and our recycling of glutathione is maxed out. We just can't detoxify fast enough."

At least not without taking action. The good news is that you can fight back—that, of course, is the premise of this whole book. Adopting the right practices can help you bring your glutathione reserves back up, if not exactly to the level you had as a middle schooler, but certainly enough to help you resist disease and make a difference in your overall well-being. But before I get to the specifics of how to increase your glutathione levels, let's talk more about what happens to make those levels go down in the first place. The more knowledge you have about how glutathione works in the body, the more the action plans in Part III of this book will make sense.

DOUBLE TROUBLE

The human body is primarily made up of water—we're about 60 percent H_2O. Glutathione is the second most plentiful molecule in the body, and it's present in all our cells. It's especially concentrated in the liver, the organ that, with the help of GSH, disarms harmful substances or makes certain that they're escorted off the premises.

Most of us start out life with copious amounts of glutathione and it builds from there (there are a few exceptions—see "Born Without Enough Glutathione," page 20). In a 2001 study conducted at a Turkish university, researchers looked at age-related changes in glutathione and found that babies have plenty of GSH and that it even begins to build starting at age two. The rise continues but, then, as the old adage goes, what goes up . . . and down it does indeed come.

The most reliable way to see how much glutathione declines from birth to old age would be to do a longitudinal study. That would involve following individuals for an extended period of time and measuring their GSH at various intervals. But that kind of study would take eons (in research years, at least) so the Turkish researchers, who are biochemists from the Osmangazi University medical school, did the next best thing, which was to look at the glutathione levels of various people in various age ranges.

The study involved taking blood samples from 176 individuals, both males and females, between the ages of 2 months and 69 years old. The 176 were divided into five groups by age: 2 months to 1 year, 2 to 11 years, 12 to 24 years, 25 to 40 years, and 41 to 69 years. Which age range do you think had the highest GSH levels? If you guessed ages 12 to 24, you're right. After age 24, our glutathione levels slip until they're lower than they were when we were developing infants. The oldest group, ages 41 to 69, had levels about 20 percent lower than the people in the 25–40 age range.

Many things, of course, change as we age so it's not necessarily a big surprise that glutathione changes, too. But if we understand *why* it drops, we can gain some insight into how to bring it back up. So, what's going on?

It's what I would call a double whammy. As you get older, the GSH manufacturing process experiences a slowdown so your body makes less glutathione. That's challenge number one. At the same time, your body's demand for glutathione rises. That's problem number two, and together with challenge number one, it creates a crisis of supply and demand.

Here's one way to think about it. Say you operate a bakery that makes the most delicious cookie in the world. It's so delicious that everyone in town wants one, causing a long line out the door. But, no problem, your bakery is equipped to serve a crowd and stays open until everyone in line gets a cookie. One day, though, you get notified that your flour supplier is experiencing a shortage and will only be able to deliver thirty pounds of flour instead of the usual fifty. What's more, one of your mixers broke. Your bakery can still make cookies, just not as many. That's a problem because now, not only do you still have the same line of townspeople out the door, but residents of the neighboring village who heard about the marvelous cookie have also gotten into the queue. Less supply, more demand. Some hungry people are not going to get a cookie.

BORN WITHOUT ENOUGH GLUTATHIONE

THERE'S A GENETIC CONDITION called glutathione synthetase deficiency, which inhibits the body from making glutathione. The disorder can present from mild to severe and somewhere in between with such symptoms as anemia, and in severe cases, neurological symptoms, such as seizures. The condition is rare but can be devastating.

Glutathione synthetase deficiency is caused by a mutation on the GSS gene, but there are also other mutations that can inhibit glutathione production. Over a year ago, I was visited in my lab by a woman named Amie, who lives in Hawaii. Amie has a condition related to a mutation of the MTHFR gene, which hinders her glutathione-making machinery. After experiencing severe digestive issues and chronic inflammation for many years—her problems were labeled as everything from reflux disease to fibromyalgia and chronic fatigue syndrome—she knew she had to make major lifestyle changes to be able to regain her health. At the time, she didn't know anything about her genetic mutation.

Amie started by reading everything she could find about her symptoms, then found a practitioner who could help her get better. Her physician diagnosed her with a severe case of leaky gut syndrome (see page 61 for more about this condition) and helped her change her diet. She began avoiding wheat, gluten, dairy, soy, and several other foods, plus added daily supplements to her diet.

Things got better. Amie's health slowly improved over the next few years, but she still tired easily, and experienced hair loss and brittle nails. Her skin was dry and showing premature signs of aging. Her naturopath suggested IV glutathione, which worked wonders for improving her energy levels, but the effects were always short-lived. (That's because IV glutathione only lasts in your body for up to three hours at a time—see page 128 for more on IV GSH.) That's when she began looking into another glutathione solution, and our paths crossed.

While visiting California, Amie stopped by a pharmacy that offered our topical GSH solution. "I was hoping for a miracle," she later told me. The effects on her energy level were significant. "I felt like the Eveready Bunny!" says Amie. Over the course of the next few months, she also noticed a change in her skin and her hair returned to normal. Suddenly, Amie's body was getting the glutathione it had been unable to make on its own. She was so excited about the changes that she flew back to California to meet with me. We went over the best way for her to use supplemental glutathione to treat her deficiency, and talked about the foods she could eat to increase her body's natural GSH production. I never met Amie before she began treating her GSH-related condition, but I can say that she certainly is vibrant now!

·······●·······

LESS SUPPLY

As you might remember from Chapter 1, three amino acids—glutamic acid, cysteine, and glycine—are combined to make glutathione. This feat of molecular engineering is dependent on a few factors. One of them is enzymes, proteins that set off biochemical reactions. Glutathione production depends on an enzyme called glutamate cysteine ligase (GCL) and, while scientists are still trying to work out the specifics, it appears that the GCL declines with age. That is likely one reason glutathione supplies in the body decline—you can't make a lot of cookies when you don't have enough mixers to stir up the ingredients.

Or enough of the primary ingredients. It may be, too, that glutathione production shrinks because there's not enough cysteine available to make it. Of the three building blocks of glutathione, cysteine is the one that's hardest to come by. And, while all three amino acids must be on hand before a GSH molecule can be formed, cysteine helps the body form glutathione rapidly—and it has a particularly important role to play in giving glutathione its detoxing power. One of the properties of cysteine is sulfur and sulfur is a sticky (and smelly) molecule. Things we come into contact with every day but don't want in our body—such as chemicals from cleaners and heavy metals hiding in food—stick to sulfur and are then kindly shown the door.

We get most of our cysteine from what we eat, so a diet lacking in cysteine-rich foods can lead to lowered glutathione levels. There is also some evidence that the mineral selenium is integral in determining how much cysteine is available for glutathione manufacturing. Researchers have found that selenium deficiencies decrease cysteine levels and lower the rate of GSH production. While how abundant your diet is in cysteine- and selenium-rich foods isn't necessarily related to age, older people, for a variety of metabolic reasons, tend to have lowered selenium levels.

Each process in the human body, glutathione manufacturing included, depends on a chain of interactions and reactions. The potential chinks I've just listed—not enough GCL, cysteine, or selenium—have the potential to gum up how GSH works to some degree, that's for certain. What's still unclear is whether these factors decrease glutathione levels to the point of deficiency, or whether, as we age, our needs increase to a point where no matter how much we produce it's not enough. Let's take a look at the demand side of GSH economics.

MORE DEMAND

In its capacity as an antioxidant, glutathione has the unusual ability to recycle itself, which I talked about on page 11. But in its role as a detoxifier in the liver, things are different. Glutathione must bind with a toxic molecule to render the toxin harmless. That changes its molecular structure, preventing glutathione from recycling itself the way it does when it's working to neutralize free radicals. So, in a heroic act of sacrifice, detoxifying glutathione goes to its death.

When you have plenty of glutathione on hand, this isn't a problem. As a seven-year-old, you may produce and be exposed to plenty of free radicals, but your body can get rid of every single one. There's no imbalance. But as early as in your twenties, with GSH production in decline and demand up (or even at the same level), your body's ability to fight free radicals recedes.

What's more, when there's an especially high demand for glutathione, the body starts sending out an SOS, repeatedly asking the GSH manufacturing system to kick into gear and make more. But like a recalcitrant teenager who's been told to clean up his room one too many times, the system says, "Stop telling me! I know you need it, but this is the best I can do!" Basically, you stop taking advice from your own body to produce glutathione. It's a negative feedback loop.

Why does the demand for glutathione increase as you age? There are several reasons. No matter how old you are (but assuming you are an adult), the world is probably more fraught with chemicals and other hazards than when you were a child. In a report from the University of California, San Francisco, Tracey Woodruff, PhD, MPH, a former senior scientist at the Environmental Protection Agency and director of the university's Program on Reproductive Health and the Environment, was quoted as saying, "In the last 50 years, we have seen a dramatic increase in chemical production in the United States." In many cities and outlying areas, air pollution exposes us to carbon, lead, nitrogen oxides, ozone, soot, and hundreds of other air pollutants on a daily basis. What's more, NASA scientists analyzing satellite data found that the amount of UV radiation reaching the earth has increased exponentially in just the past thirty years.

What you're knowingly putting in your body can also affect the demand for glutathione. If, like many people as they age, you take a variety of medications, your system will order up extra GSH to help it neutralize and eliminate foreign compounds in the drugs. Even acetaminophen (Tylenol), a seemingly harmless painkiller, has been associated with depleting levels of glutathione. When you take acetaminophen, your liver produces small amounts of a compound called NAPQI (N-acetyl-p-benzoquinone). Glutathione's job is to prevent NAPQI from doing harm, but take acetaminophen for an extended period of time, or too much at one time, and it puts GSH production into overdrive, until ultimately you run out. (I'll talk more about acetaminophen and how it figures into opioid overdose on page 171.)

Then there's what you eat. If you primarily dine on processed foods, you also regularly consume chemicals designed to preserve, color, flavor, and texturize what you eat. Even if you're a big vegetable eater, pesticides can be clinging to the leaves of your green beans and broccoli. Again, glutathione will be called on to get rid of those potentially harmful substances. Alcohol can pose another problem. In mice studies, alcohol has been shown to exhaust glutathione reserves, especially in older animals. This can be particularly debilitating since the body relies on GSH to help the body recover from alcohol intake.

This laundry list of ills isn't meant to make you feel as though you're a walking hazardous waste site — the body does a pretty good job of removing most harmful substances, even when glutathione is in decline. But it's important to be aware that many chemicals and heavy metals are fat soluble, which means that they can get stored away in fatty tissue. For that reason, it's possible to accumulate more and more of these substances in your body where, over time, they can help accelerate aging and cause disease.

Demand for glutathione also increases when there's chronic inflammation in the body. The inflammatory process creates oxidative stress, and the more oxidative stress you have, the more glutathione needed to staunch the increase in free radical activity. One common source of chronic inflammation is diabetes, a disease we're seeing in epic proportions. According to the Centers for Disease Control and Prevention (reporting in 2017), more than 100 million adults in the US have diabetes or prediabetes — that means there's a lot of people with inflammation out there.

And a lot of people who have low glutathione levels. Several studies show that there's a direct correlation between type 2 diabetes and diminished glutathione. In 2011, researchers writing in *Diabetes Care* found that not only did people with diabetes have lower GSH levels than control subjects without diabetes, they also synthesized glutathione at a 43 percent slower rate.

INFLAMMATION: THE GOOD, THE BAD, AND THE UGLY

THE HUMAN BODY IS a wondrous machine, so intricate and intelligent. But it's not infallible and sometimes, as in the case of the inflammatory response, the body's best intentions go wrong.

Inflammation is a by-product of the immune system's response to insult and injury. Think of what happens when you twist your ankle. It turns red and swollen (the usual definition of inflammation) not because of the injury so much as because your body's white blood cells immediately go to work to heal the damage. What happens first is a release of messenger cells that send hormones to the affected area where they cause blood vessels to dilate. This allows more healing immune cells to reach the trouble spot. More fluid is also sent to the area. Together, the two processes turn your ankle—the infected cut on your finger, your stubbed toe, or any other assaulted area—red and puffy. It happens fast and usually subsides after a short time. The same cascade of events also happens internally where we don't see it. Bronchitis, for instance, is inflammation of the bronchi (passageways in the lungs); cystitis is inflammation of the bladder. These are problems, of course, but assuming the inflammation leads to healing, this is what we consider "good" inflammation.

What about bad inflammation? Chronic inflammation is inflammation that lasts for months or even years, although it's so low grade you may not even know it's there. This type of inflammation can occur when the offending agent, say bacteria or a tissue-aggravating chemical you breathed in or ingested, isn't entirely eliminated by the body. It can also occur when the immune system doesn't turn off the response per normal operating procedure. Either way, the result can be damage to tissues in various areas of the body with repercussions that range from small to very large. According to the National Institutes of Health, common symptoms of chronic inflammation include body pain; constant fatigue and insomnia; depression, anxiety, and mood disorders; gastrointestinal complications, such as constipation, diarrhea, and acid reflux; weight gain; and frequent infections.

Where it can get really ugly is when chronic inflammation leads to disease. Several disorders, in fact, are categorized as chronic inflammatory diseases, including rheumatoid arthritis, inflammatory bowel disease, and multiple sclerosis. Chronic inflammation has also been implicated in cardiovascular disease, cancer, and diabetes among other conditions.

Sometimes the factor behind or contributing to chronic inflammation is oxidative stress. Free radicals can stimulate an inflammatory response—and the inflammatory response can also stimulate the production of free radicals. Whether this tit-for-tat process leads to trouble can depend on the balance of antioxidants in the body and most especially glutathione. Some research, for instance, has shown that GSH protects against inflammation in the lungs due to lung disease.

What can you do if you suspect you have chronic inflammation? The first line of defense is to eat healthfully by maintaining a diet that is rich in fruits and vegetables. Highly processed carbohydrates have been linked to inflammation, as has red meat, so keep them to a minimum. Keeping your body weight within the healthy range can help too (obesity is associated with inflammation). If you continue to be concerned, get tested.

A few tests provide clues to inflammation. One of them is the CRP test that measures C-reactive protein, a marker of inflammation. There is also the high-sensitivity CRP test, which, as the name suggests, is a more sensitive test. It's often given to people at high risk for heart disease. Another indicator of inflammation is a homocysteine test. This amino acid reveals deficiencies in vitamins B_6, B_{12}, and B_9 (folic acid), which can also be related to inflammation. Talk to your doctor about these tests, then have them advise you on next steps according to the results.

········•········

HOW DO YOU KNOW IF YOU HAVE LOW GLUTATHIONE? SHOULD YOU HAVE YOUR GSH LEVELS TESTED?

IT WOULD BE WONDERFUL if we could just look at a list of symptoms and use it to definitively determine what's going on inside someone's body. Unfortunately, medicine rarely works that way. It follows, then, that I can't tell you the exact signs of GSH depletion. In fact, since glutathione is involved in so many biological actions, almost everything that can come up—from fatigue to stomach problems to weight gain to dull skin—can be a symptom of low glutathione. I know—that's not very helpful!

But let me say this: If you are not feeling your best or have been struggling with an intractable illness, there's a good chance that improving your glutathione levels can help. As you read through this book, I think the information you take in as well as certain stories of individuals who struggled with symptoms before treatment with GSH will resonate with you. Those aha! moments may lead you to see how bumping up your own glutathione level through one of the Action Plans in this book works. Trial and error with no risk, since each plan is safe and healthy.

There is also another course of action: Have your glutathione levels tested. It's possible to do so through a simple blood test and it's certainly an option, although not as perfect as it might seem.

Theoretically, knowing the glutathione state of affairs in your body can provide clues to whether your body has disease or if the antioxidant/detoxifier is being depleted because of medication you're taking or some other reason. Yet there's one problem inherent in testing glutathione (and, similarly, with testing many vitamins and minerals): Glutathione use fluctuates throughout the day and varies according to what demands are being made on the system at a given time. If you had a few glasses of wine the night before you have your glutathione test, your results may be different than if you drank water with your meal. If you recently took a pain reliever, are fighting a virus, or just spent an hour exercising—all these factors could affect your test. It might take several rounds of tests to detect your average level of glutathione.

A glutathione test can also miss glutathione that's in the recycling stage. You might remember that after two glutathione molecules each donate an electron to a marauding free radical, they team up to form a new molecule called GSSG. Eventually GSSG will grab some new electrons, split, and be recycled as two new GSH molecules. But a test won't catch the GSSG form of glutathione, potentially making it seem as if your levels are lower than they really are.

Glutathione tests are generally administered in doctors' offices and can cost upward of $125. Is it worth the money? If you control for certain factors there may be value in getting a glutathione test. If you have a condition you and your physician are considering treating with glutathione, then a test may give you a starting number, whereas later retests can help you see whether the treatment is working. There's no right or wrong here; just make sure you consider all the factors I mentioned before opening up your wallet.

POSSIBLE CAUSES OF LOW GLUTATHIONE LEVELS

- ▷ Aging
- ▷ Genetic deficiency
- ▷ Illness
- ▷ Chronic inflammation
- ▷ Selenium deficiency
- ▷ Cysteine deficiency
- ▷ A surplus of toxins in the body
- ▷ Autism

THE EFFECTS OF LOW GSH

You already know that glutathione is the most important antioxidant in your body and that it plays a critical role as a detoxifier as well. But glutathione is also involved in a number of other processes inside cells, including acting as a chemical messenger, regulating the life and death of cells, and helping maintain proper energy production in the mitochondria. Mitochondria are tiny factories inside your cells that take food and oxygen, digest it, then turn it into energy for your body. They're like little Pac-Man figures that munch away, digesting nutrients, and spitting out energy in return. Since mitochondria are the "engines" of your body, you want to take good care of them. And one of the ways to support them is with adequate levels of GSH. When glutathione is unavailable to help out the mitochondria or fulfill any one of the other tasks I mentioned, it can have detrimental effects on your health.

I've already told you a bit about the connection between glutathione and diabetes, and in the next section of the book I'll detail other links between glutathione and disease as well as premature aging. But I want to begin the discussion here by talking about some of what we know about the consequences of dips in GSH. In broad strokes, we can say for certain that people with heart disease, arthritis, diabetes, cancer, and neurological diseases, such as Parkinson's—to name just a few adverse conditions—have lower levels of GSH than control subjects. This suggests that GSH plays a role in protecting against disease and that, when it goes missing, disease onset may follow.

But does low glutathione cause disease? "Cause" may not be the right word. It's more like what happens when you take a loaf of bread out of its wrapper. Removing the wrapper doesn't *make* the bread turn moldy, but it allows it to happen more rapidly. Likewise, glutathione going missing doesn't cause cancer or diabetes or Alzheimer's, but its absence likely opens the door to damage that does cause these diseases.

How exactly? That's still under investigation. Glutathione plays so many roles in the body that scientists aren't sure exactly why low levels are associated with disease. But they do have some ideas.

In Japan, researchers have been conducting a population-based study in the suburban community of Hisayama since 1961. In one particular part of the study, the researchers measured the glutathione levels of 176 people who had previously suffered a stroke or heart attack. They then compared the results to blood samples from healthy individuals. As in so many GSH studies, glutathione levels were much lower in the people with disease than in those without, particularly when it came to people who'd had cerebral hemorrhages.

The nature of the study didn't allow researchers to pinpoint how glutathione protects against heart attack and stroke, but they did note that certain areas of the brain are more sensitive to fluctuations in glutathione, one potential reason low GSH is linked to stroke. They also hypothesized that, since glutathione protects against both oxidative stress and the oxidation of LDL cholesterol—two risk factors for heart attack—its absence may set the stage for cardiovascular disease.

There have been thousands of studies on glutathione, but clearly we still have a lot to learn. Thankfully, recognition of how integral GSH is to good health is growing and I expect the result will be broader awareness of how important it is to adopt lifestyle habits that maximize glutathione production in the body. More research will also help test the parameters of using glutathione in not just the prevention but the treatment of disease. In fact, we're already starting to see some applications of glutathione in the fight against diabetes and other conditions. I'll share more about that later in this book.

BEYOND THE BODY: DOES LACK OF GSH AFFECT THE MIND?

As we've learned more about mental health over the years, the clearer it's become that psychological states of mind can have biochemical origins. The connection between messaging chemicals called neurotransmitters and depression, for instance, is quite clear. Neurotransmitters as well as hormones can be involved in the development of anxiety disorders too. Less studied, but still significant, is the fact that some forms of depression and anxiety may be linked to oxidative stress. And one way that scientists have determined this is by discovering the presence of low glutathione levels in people (and, in some cases, animals) suffering from these psychological conditions.

For example, in 2017, researchers from the Icahn School of Medicine at Mount Sinai and Weill Medical College of Cornell University joined together to publish a paper on adolescent depression. Using magnetic spectroscopy (a magnetic resonance test that looks at biochemical changes in the brain), the group compared the glutathione levels of adolescents suffering from depression with those of healthy control subjects. They found that the depressed adolescents had significantly lower levels of GSH. Other studies have also shown that glutathione is in shorter supply in depressed individuals as well as in animals exhibiting anxiety-like behavior.

When researchers look at glutathione levels in these studies, the question they are often asking is, Does oxidative stress play a role in this particular disorder? In other words, glutathione is a *marker* of oxidative stress (which you know by now is the unfavorable ratio of antioxidants to free radicals). If glutathione is low, it indicates that it's been used up fighting free radicals. To me, the more salient question is, If more glutathione was available, would the problem be averted? That's something we don't know the answer to right now, though I hope it will be addressed in the future.

Many different factors can be involved in depression and anxiety. Among them is dysfunction in the brain chemicals serotonin and dopamine. Some researchers believe that glutathione helps dopamine in the brain become more effective and increases sensitivity to serotonin. If this is the case—and it still hasn't been proven yet—then glutathione may one day be used to help treat depression. Right now, I don't know of many doctors treating either anxiety or depression with GSH. However, as we learn more about the relationship between oxidative stress and psychological disorders, we may find that glutathione has an important role to play.

While we're on the topic of states of mind, it seems a good time to ask: Are glutathione and psychological stress—i.e., the stress you experience over such things as fluctuating finances, marital woes, and professional pressures—at all linked? The answer is no—but, indirectly, also yes.

There are basically three types of stress we humans experience. There's physical stress that occurs when you get ill or injured or the body is pushed to its natural limits. There's chemical stress, which is a by-product of exposure or ingestion of some type of toxin or pollutant. And there is psychological stress, the consequence of anything that weighs on our mind. Glutathione is primarily involved with chemical stressors. Its job is to get rid of anything harmful in the body, be it heavy metals or free radicals. Psychological stressors are mostly handled by the adrenal glands, which release the hormone cortisol to help your body cope with anything your brain sees as a threat. So, in that sense, glutathione isn't called upon to assist you in dealing with road rage when someone cuts you off in traffic or the tension you feel when you've got a big project due at work. Psychological stress doesn't "use up" GSH.

However, reducing stress in your life can help your body get the most out of glutathione. Here's why: Among the many adverse effects stress has on the body is that it raises the heart rate and interferes with how deeply you breathe. These are two things that can inhibit glutathione's job of detoxifying the body. One of the reasons I've included a short meditation in Action Plan One is to provide a way to help you slow your heart rate and breathe deeply. This will increase the oxygen in your body and allow you to have a better exchange of gases through the lungs, both of which will give GSH an assist in the detoxification process.

GSH AND AUTISM

Luckily, most people are born with fully operational glutathione-making machinery. But there is a small subset of people who have GSH levels 20 to 40 percent lower than their peers: children with autism. Autistic kids have also been found to have other metabolic abnormalities, including low levels of cysteine, that all-important component of glutathione. Another thing we know about children with autism is that many of them have very high levels of oxidative stress, a condition most likely connected to their lack of glutathione. The question is, Is oxidative stress—and, by extension, the lack of glutathione that fails to counter it—responsible for some of the symptoms of autism?

Autism and the spectrum of disorders that fall under the autism umbrella are generally characterized by challenges with social skills, behavior, and communication. The Centers for Disease Control and Prevention estimates that one in fifty-nine children in the United States are on the spectrum. Autism is thought to develop through a combination of genetic and environmental factors, although the exact cause has evaded researchers so far.

Aside from the usual social and behavioral problems faced by children with autism, many autistic kids also have gastrointestinal problems—about 41 percent, according to a 2011 study published in the *Journal of Developmental & Behavioral Pediatrics*. With this statistic in mind, a group of researchers from Children's Hospital in Los Angeles, University of Southern California, and Vanderbilt University measured a marker of oxidative stress in four groups of children. Group one had only autism, but no gastrointestinal symptoms; group two had only had gastrointestinal (GI) symptoms, such as diarrhea, constipation, and reflux; group three had both; and group four had neither. All of the children with either autism, GI symptoms, or both had elevated levels of oxidative stress, but the kids with the highest levels were those whose autism was accompanied by GI dysfunction. Studies like this one can't show direct cause and effect, but nonetheless it does indicate the likelihood that oxidative stress contributes to the gastrointestinal problems experienced by many autistic kids.

There's also been some evidence that points toward the involvement of oxidative stress and glutathione impairment in other symptoms of autism. Over the years, researchers at the University of Arkansas have conducted many studies looking at glutathione, oxidative stress, and autism. In 2012, they published findings of decreased GSH activity and free radical damage in regions of the brain associated with behavior.

Then, also in 2012, they looked into whether boosting levels of glutathione would affect some of the symptoms of autism. This particular study involved thirty-seven children who had both autism and abnormally low levels of glutathione. For three months, the children followed a supplementation regimen with methylcobalamin (a form of B_{12}) and folinic acid, two compounds known to improve GSH synthesis. Their glutathione levels did indeed go up—and so did their scores on a scale that measures personal and social skills. As reported by parents, during the three months of supplementation, all the children had improvements in expressive communication, personal and domestic daily living skills, and interpersonal, play-leisure, and coping social skills. This is a bright spot in the quest to treat kids on the spectrum and, hopefully, it will spur more research into the glutathione-autism connection.

AN AUTISM CASE STUDY

WE ARE JUST BEGINNING to hear about some very exciting anecdotal evidence of glutathione's power to help children with autism. One story is so affecting I'd like you to read it in the child's mother's own words. Her name is Alyssa and here is what happened:

> Lights of cars driving by flashed into the minivan windows. It was one in the morning at the Coco Keys resort parking lot and I sat in the dark, full of fear, frustration,

and defeat as my 2-year-old daughter threw another epic tantrum. I imagined what I would say to the police officers if they happened to walk by and saw my toddler banging her head and hands on the car windows.

My daughter was born with a frown on her face and even as an infant she did not take naps longer than 45 minutes. She rarely held eye contact with anyone and seemed to be locked inside her own world. She had trouble falling and staying asleep. We tried sleep training her and on the worst night she screamed and cried for five hours straight until we gave in and got her out of bed. During this time, she made bite marks all over her crib and pulled out all her hair. Family members and friends were expressing concern and we finally took their advice and had my daughter tested. At 28 months, she was diagnosed with autism spectrum disorder (ASD). Suddenly so many things made sense.

As parents of an ASD child, my husband and I made it our mission to do research and adopt any interventions that could help our daughter manage and eventually overcome her condition. We started intervention therapy early on and, while she improved, change came at an excruciatingly slow pace. Her tantrums continued and she was still a very unhappy and unpleasant little girl. At the age of two, she had still not spoken a single word.

So needless to say, we were ready to try anything when a friend recommended that we research topical glutathione. With little to no expectation, I began to spray my daughter twice a day. I would have to say that the positive effects came almost immediately. Within two weeks our daughter's behavior significantly improved. Before the GSH, she was having major 20 to 30 min tantrums several times a day—but after using glutathione, her tantrums became shorter and less frequent. Even the therapist treating her noticed a significant drop in inappropriate behavior and said it was miraculous how quickly she'd improved. Not only did her moods get better, but she slept longer and woke up less frequently. Before the glutathione, she wasn't sleeping nearly enough, so with the improved sleep, she was also less irritable and a more pleasant child.

As a mom, the most exciting thing was witnessing my child's mind waking up and starting to connect with the world around her. But most amazingly, within days after she began detoxing with the GSH she uttered her first word, "hi," to a friend at church. I burst into tears at the shock of hearing her voice. After this word more words followed and she even began experimenting with new sounds. She's still "delayed" compared to other kids her age, but the improvements she has made after the glutathione are dramatic and significant.

········●········

PART II

Protector, Detoxifier,
Energizer, Beautifier

3 DEFENSE AGAINST DISEASE

disease (dis • ease, di-'zēz), *noun*

1. A disorder of structure or function in a human, animal, or plant, especially one that produces specific signs or symptoms or that affects a specific location and is not simply a direct result of physical injury.

That's what the *Oxford English Dictionary* has to say about disease. But let me add another definition, one that I sometimes use to highlight the mind-body connection.

2. Lack of harmony in the body.

One of my favorite quotes from Sri Aurobindo—who is my family's revered hometown guru from Pondicherry, India—is this: *All problems of existence are essentially problems of harmony.* This is particularly true when it comes to our health. When there is disharmony in any system or ecosystem, it will eventually lead to a breakdown and, in some cases, complete chaos. If you look around the world today, there is a severe lack of harmony and a lack of peace in our societal ecosystem causing a lack of ease—that is, dis • ease—all over the globe.

When it comes to your individual body, the same is true. If the ecosystem of your body is not harmonic, it will be out of balance, opening up the door for you to become ill. When disease shows itself to you, usually in the form of symptoms that make you feel sick and perhaps send you to your doctor, it is functioning as an alarm, trying to get your attention. Disease is your body screaming at you, asking you to find the root of the disharmony. You will get this same sort of alert if your glutathione supply and demand are out of balance. This imbalance will make itself known through one, or quite possibly many, uncomfortable symptoms. These symptoms are an indicator that something below the surface is going wrong.

Say, for example, that one day you wake up with a rash covering your body. Whether that rash is from an allergic reaction or something like a bug bite, it acts as bread crumbs that can be followed back to the source of disharmony in your body. Hopefully, the crumbs will lead to the source of the problem before whatever is going on worsens and turns into a full-fledged disease.

I say all of this to highlight and reiterate what I will express several times in this book: Our body is designed to fight for us, and to heal itself. Our body is wired to seek "ease," and we are meant to feel good and function at optimal levels. As a society, as stress levels go up and environmental toxins multiply, we have learned to accept discomfort and disease as if it is just a normal part of being human. I would like to raise my hand here (and shout from the rooftops, if I may) and say that we need to stop accepting discomfort and disease as part of life. When we are not feeling good, and something feels off, we need to train ourselves to look for the early warning signs of disease. I personally do not accept anything less than feeling energetic and fully functioning every day.

This thinking underscores the medical evidence that most disease states are associated with *low* levels of glutathione and that people with long life spans have *high* levels of GSH. In the previous chapter, I introduced you to some of the diseases associated with low glutathione levels. Remarkably, numerous other maladies can develop in the absence of adequate glutathione.

And as this connection becomes clearer, everyone from university-based researchers to pharmacists like me are looking into the prospect of using glutathione—or GSH precursor compounds known to kick-start the body's own GSH production—to treat these diseases.

Unbeknownst to the majority of people, our bodies actively fight off disease every day. And, for the most part, they are very efficient machines. But, of course, sometimes they fail and disease occurs. Different diseases develop for different reasons; however, there are some commonalities, and lack of glutathione is one of them. I'm going to give you a general overview of how glutathione levels are linked to disease causation, then we'll take a look at some of the common maladies that can be prevented and/or remedied by raising glutathione back up.

DISEASES ASSOCIATED WITH LOW GLUTATHIONE LEVELS

- Asthma
- Autoimmune disorders (rheumatoid arthritis, lupus, psoriasis, contact dermatitis, Crohn's disease)
- Cancer
- Cardiovascular diseases (atherosclerosis, coronary artery disease, arrhythmia, cardiomyopathy)
- Chronic obstructive pulmonary disease (COPD)
- Cystic fibrosis
- Diabetes
- Diabetic retinopathy
- Fibromyalgia
- Macular degeneration
- Neurodegenerative disorders (Alzheimer's, Parkinson's, ALS, Huntington's)
- Stroke
- Viral diseases (common flu, HIV/AIDS, herpes, shingles)

THE ITCHY TAG AND YOUR IMMUNE SYSTEM

A FRIEND OF MINE was attending a wedding with his brother, who kept complaining about the tag on the inside of his shirt. The guy must have really loved this shirt because he was wearing it often enough to have developed a rash from the paper tag rubbing against his side. He tried cutting off the tag, but the remaining bits irritated his skin even more.

My friend happened to have some of the Auro GSH produced by my company and suggested that his brother apply some of it to the rash. He did—and the rash went away that very same day. Disappeared. Once the rash was gone, my friend's brother discovered that the tag had actually caused more than simple irritation; there was bruising, too. More glutathione was applied topically, and the bruising also vanished in a short amount of time.

The lousy tag on my friend's brother's shirt had caused an immune response with results that were apparent to the naked eye. He could see it, then treat it accordingly. But what happens when your body is bombarded with things known to cause harm—such as elements in foods you may not even know you're sensitive to, or pesticides, or pollutants from the air, or even free radicals developed by your own exercise exertion—and you can't see the damage? Like the irritating tag, these attackers may trigger an immune response in your body, but you won't have a clue as to whether your immune system is successfully thwarting damage.

What's more, the response itself, by creating inflammation, can also do harm to the body in other ways. By the time you feel the symptoms, you may already be in trouble. Which brings me back to glutathione. Wouldn't it be wonderful to have a master antioxidant already on the job protecting against potential invaders? A body with plenty of GSH is a body with an insurance policy, all the more reason to follow the Action Plans beginning on page 133.

OXIDATIVE STRESS ON THE LOOSE

One of the conditions that allows disease to take hold is oxidative stress. Oxidative stress, as noted in earlier chapters, occurs when there is an imbalance between free radicals and the antioxidants needed to stabilize them. If they're not neutralized by antioxidants, free radicals will snatch electrons from other molecules—a process known as oxidation—leaving those "robbed" molecules stressed and unstable. This, in turn, can set the stage for both chronic and degenerative diseases.

One way that oxidative stress inflicts harm is by messing with your lipids. Lipids are fats in the body that make up cell membranes, help you store energy, and act as chemical messengers, to name just a few of their jobs. When lipids come under attack by free radicals, the cells they inhabit can become structurally damaged and even die. Fat oxidation doesn't only happen in the human body. Have you ever heard the recommendation that you store olive oil (a lipid just like the fat in your body) in a cool, dry, and dark place, and preferably in a dark container? This advice is based on the fact that light causes olive oil to *oxidize*—or as you'll experience it the next time you try to mix it into a salad dressing, cloudy and rancid. You don't want this to happen to your high-priced olive oil, let alone your precious body!

Oxidative stress can also take a toll on proteins, inhibiting the all-important enzyme activities that catalyze chemical reactions in the body. Perhaps most damaging of all, oxidative stress can harm DNA, the helix on which your genetic material is encoded. DNA functions like command control, providing instructions for the reproduction, growth, and behavior of every cell and, by extension, your whole body. But DNA, while it may be critical to your very existence, is also very vulnerable. It can easily take hits from free radicals, diminishing its ability to repair itself and leaving it prone to unhealthy reconfigurations. One of the duties of the genes is to alert the immune system when it needs to step up and suppress a tumor cell or virus. When DNA is altered, the genes' signaling ability may be impaired—they can no longer sound the alarm when needed.

How these potential scenarios play out can vary. But we know for sure that they can be at the genesis of some of the most common and serious diseases. Luckily, when you're young and glutathione is plentiful, you don't have to worry too much about oxidative stress and its consequences. But as you get older, your body isn't as well equipped to handle potential challenges. One way to look at it is through this front door–back door parable:

You're at home, relaxing on the sofa in your living room, when a neighbor comes to your front door and rings the bell. At the same time, another neighbor begins knocking on your back door. Which door do you answer?

You know the neighbor at the front door well. He comes over nearly every day, and you know that when you invite him in and, in a neighborly fashion, offer him some refreshments, he practically eats you out of house and home and talks your ear off as well. But you know this about him and are prepared. You've got a strategy to make sure he gets a cookie and a cup of tea and is out the door in ten minutes. Now, the neighbor at the back door is unfamiliar to you. When you let her in, she may start snooping around (you've heard she's nosy) and peppering you with so many questions that it takes up a great deal of your time.

So, let's say that both neighbors don't even wait for you to answer the door. They just go ahead and let themselves in. When they enter the house at the same time, you make the decision to greet the new neighbor to ensure that she doesn't snoop too much or hang around too long. And since she's unfamiliar to you, it takes some effort to rein her in. Meanwhile, as you're dealing with the neighbor who came through the back door, the front door is wide open, allowing the more familiar neighbor to saunter in and say to himself, "You know what, no one is here to tell me what to do and not to do. I'm going to poke around." Suddenly he's pulling out the peanut butter and jelly, pouring himself a beer, and switching on Netflix. Chaos abounds.

Oxidative stress is the familiar neighbor who comes to the front door. It's a daily, ongoing issue that your body deals with regularly. You've got the tools—including natural killer cells, a part of the immune system that is dependent on glutathione—to not only nip oxidative stress in the bud, but also wipe out any potentially cancerous cells, viruses, and other threats that slip through the cracks. But what happens when unforeseen, back-door issues crop up? If you're unexpectedly dealing with exposure to sunshine, pollution, heavy metals, pesticides in food, and other abnormal sources of oxidative stress, the front-door issues—issues you can normally handle— have the potential to spiral out of control.

That is one way to think about how disease sneaks in when glutathione supplies are overtaxed. But individual illnesses also have their own unique ways of developing, and thanks to researchers around the world, we're getting a clearer understanding of how the antioxidant fits into each individual picture. Here's what we know so far about glutathione and some of the common diseases we're all hoping to avoid.

ALZHEIMER'S AND OTHER NEURODEGENERATIVE DISEASES

When you're young, it's easy to joke about having Alzheimer's if you forget someone's name or where you left your shoes. But with age, the laughter kind of goes out of the wisecrack since we know that the risk of neurodegenerative diseases, such as Alzheimer's and Parkinson's (the second most common neurodegenerative disorder), goes up dramatically as you grow older. In regard to Alzheimer's, 1 in 10 individuals older than age 65 has the disease; at age 95, that number increases to 5 in 10. By one estimate, nearly six million Americans have Alzheimer's disease and other forms of dementia, and at least a half million are living with Parkinson's disease. The threat, in other words, is very real.

Neurodegenerative diseases are conditions that involve the loss of function or death of nerve cells in the brain or peripheral nervous system. Besides Alzheimer's and Parkinson's disease, some of the disorders you might be familiar with are amyotrophic lateral sclerosis (ALS) and Huntington's disease. The cause of and cure for these conditions have been the holy grail of medicine for some time and, while nothing is definitive, we are learning more every year. One thing we know is that a combination of genetics and environment is involved in the development of all these neurodegenerative conditions. You may, for instance, be genetically inclined to develop Alzheimer's, but whether you ever do may depend on what you are exposed to during your lifetime. Everything from the air you breathe to the foods you eat may matter.

What science also tells us is that glutathione is one of the weapons the brain utilizes to guard against the kind of oxidative damage to DNA and proteins that has been found in patients with neurodegenerative diseases. Accordingly, there is a very strong link between low glutathione levels and these diseases. Two studies, for instance, have found that people with Alzheimer's disease and even mild cognitive impairment have significantly lower GSH in both the hippocampus and frontal cortex areas of the brain as compared to healthy people of the same age. Similarly, we have known for a long time that glutathione is considerably reduced in the brains of people with Parkinson's disease. Way back in 1991, researchers at the University of Southern California School of Pharmacy (my alma mater, as it happens) found that, in advanced cases, glutathione was only 2 percent of the normal amount. GSH depletion has also been identified in ALS, Huntington's, and other neurodegenerative diseases.

By now you will not be surprised to learn that where glutathione is low, oxidative stress is high. In a review of studies looking at oxidative stress in people suffering from Alzheimer's, researchers from Johns Hopkins School of Medicine and the National Brain Research Institute in India noted that people with the disease have elevated levels of oxidized proteins and DNA in the areas of the brain associated with dementia. And the more progressed the disease, the more oxidation present.

What these studies tell us is that glutathione in the brain is essential. While the antioxidant is everywhere in the body, it's particularly concentrated in certain areas. The liver is the organ with the highest level of glutathione, but the brain also has fairly high concentrations of GSH. And glutathione is by far the most prevalent of all the antioxidants in the brain. And that's no accident. Even though the brain is only a small part of the body, the cells of the human brain consume about 20 percent of the oxygen utilized by the body as a whole. Turning all that oxygen into energy inevitably generates a great deal of free radicals, which means that the body is going to have to call in the big guns: GSH. But if there's not enough glutathione to answer the call? Combined with genetic and environmental factors that predispose someone to a neurodegenerative disease, it may just be the perfect storm.

There have been several efforts in the past to develop a glutathione treatment for neurodegenerative disorders. Some of the most interesting Alzheimer's research going on now focuses on using compounds that increase glutathione production in the brain to help the body replenish its oxidation-fighting forces. Research in this area is ongoing and I know that I for one am going to be paying close attention to the results.

So far, researchers have had mixed results when attempting to lessen Parkinson's disease symptoms with GSH. In 2009, David Perlmutter of *Grain Brain* fame and colleagues published a study looking at the effects of IV glutathione on patients with Parkinson's. Twenty-one patients received IV glutathione three times a week for four weeks and had mild improvement in symptoms, leading the researchers to call for further study. I myself have seen glutathione help people suffering from Parkinson's. One patient of mine, named Kathy, is sixty-seven and recently retired from a forty-year career as an elementary school teacher. She is a two-time cancer survivor and was recently diagnosed with Parkinson's disease. When Kathy tried topical GSH, she saw a difference immediately in her symptoms. "I had tons more energy, and felt more alert," she told me. "I'm also sleeping better. And it definitely helped with the shakiness—almost instantly." Glutathione also helped her skin come back to life—the chemotherapy had left it dull and pasty—giving it a brightness she hadn't seen for a long time.

HEART DISEASE

"Heart disease" (also regularly called cardiovascular disease) is actually a fairly vague term. There are so many different conditions that affect the cardiovascular system, ranging from atherosclerosis (hardening of the arteries) and coronary artery disease (specifically hardening of the arteries that lead to the heart) to arrhythmias (irregular heartbeats) and cardiomyopathy (weak heart muscle). For this discussion, though, let's talk about heart disease in terms of conditions that involve narrowed or blocked blood vessels that can lead to a heart attack, chest pain (angina), or stroke.

When fatty deposits called plaque build up on the inside of the blood vessels, they slim these passages, restricting the flow of oxygen and nutrients through the body. If a piece of plaque breaks off, blocking blood flow entirely, it can cause a heart attack or stroke. Cellular damage from oxidative stress is known to play a role in clogging the arteries. Free radicals attack LDL (bad) cholesterol and other lipids, altering them in ways that cause them to clump and cling to the linings of these all-important passageways. It's as if you sent melted candle drippings down the sink and now the walls of the pipe are thick with sticky, waxy stuff. Send enough drippings down and one day the pipe will back up.

Heart disease is multifactorial—that is, there can be many factors contributing to its onset. Yet, as several studies have noted, diminished levels of glutathione seems to almost always play a role. For instance, a study measuring glutathione in the blood of seventy-six New York Heart Association patients with coronary artery disease of varying severity found that they all had lower GSH levels than healthy controls. The patients who were known to have heart disease but were asymptomatic had blood GSH levels 21 percent lower; the patients with the most severe symptoms had levels 40 percent lower. The researchers noted that glutathione drops with the gravity of the disease and that it may be a good predictor for early heart disease detection.

Glutathione, it's clear, protects the heart in some way. But low glutathione levels may be indicative of trouble not only because there isn't enough GSH to stop marauding free radicals, but also because there isn't enough glutathione to recycle the other antioxidants that also prevent oxidative damage. Glutathione, in other words, has many jobs to fill—take it out of the equation and you're in trouble.

So, what happens when you put glutathione back into the equation? Some researchers have attempted to do so through various means, including dietary glutathione supplements and intravenous methods, but found little improvement. However, in 2017, a small Italian study tested a new formulation of GSH, a pill called Oxition placed under the tongue, which is able to bypass the gastrointestinal tract and enter the circulation directly. The study involved sixteen men between the ages of forty and sixty with a variety of different risk factors for cardiovascular disease, including high blood pressure and high cholesterol. Over the course of twelve weeks, some of the men received Oxition, while others received a placebo. By the study's end, the men who took the GSH supplement had lower LDL cholesterol and less arterial stiffness as compared to the men who took the placebo. This was a small study and we can't draw any conclusions from it yet, but I found the results optimistic and hope that other heart disease researchers will also look into glutathione treatments.

My own work with glutathione has led to the question of whether transdermal forms of the antioxidant can help reduce risk factors for heart disease. And there are signs that it may. One risk factor for heart disease is high triglycerides, blood fats that can clog up the cardiovascular passageways. In a small study, my colleague Ray Solano, PharmD, a clinical pharmacist and board-certified nutritionist who practices in Cedar Park, Texas, found glutathione had a significant lowering effect on triglycerides (see page 65 for the details). Some anecdotal reports also indicate that glutathione may help with the aftereffects of heart attack or stroke. One patient, Kenneth G., an eighty-two-year-old stroke victim, began using topical glutathione to soothe a sunburn. Inexplicably, he also got a kick of energy every time his wife applied GSH. "The glutathione helped to clear my mind," he says.

DIABETES

So far, I've talked about the fact that people with type 2 diabetes have lower levels of glutathione (see page 24). Later in this chapter, I'll also discuss how research has shown the deficiency also makes them more vulnerable to tuberculosis. In fact, having type 2 diabetes makes you more vulnerable to just about *every* disease.

Type 2 diabetes occurs when the body doesn't make enough insulin, or the cells are resistant to it. (People with type 1 diabetes don't make insulin at all. This largely genetic autoimmune disease has a different pathology, so I'm going to omit it from this discussion. Anytime I refer to "diabetes," I'll be talking about type 2.) Insulin is a hormone that's essential in the food-to-energy production line in your body. After you eat carbohydrates, digestive enzymes break them down into a form of blood sugar called glucose. Insulin is then called upon to shuttle glucose into the cells where it is turned into energy. You've probably heard the term *high blood sugar*. That's what happens when insulin can't deliver glucose and it backs up into the bloodstream, creating all kinds of havoc, including, over time, damage to the eyes, kidneys, and heart.

I don't think it's a coincidence that the prevalence of type 2 diabetes has risen in tandem with the surge of junk- and processed food–eating in this country. This type of carb-heavy, salty food can increase blood fats called triglycerides, blood pressure, and blood sugar levels, all of which can set the stage for diabetes. As I mentioned earlier (and is a fact that bears repeating), according to a Centers for Disease Control and Prevention report in 2017, more than 100 million people in the US are living with prediabetes or diabetes. Recently, I read a letter to the editor in the *New York Times*, written by a family physician. His assessment says it all: "I have been a family doctor for 32 years. I used to send all diabetics to a specialist, but now every other patient is diabetic. It is a losing battle."

I hope that the last part of his letter to the *Times* isn't true and, in fact, I believe that if people changed their diets for the better (and many people are), it would go a long way toward remedying the problem of type 2 diabetes. And glutathione can help.

It's unclear why people with type 2 diabetes produce less glutathione than their healthier counterparts. Some researchers have hypothesized that it's due to disrupted protein breakdown in diabetics. Since the breakdown of proteins makes amino acids available for synthesizing glutathione, this unhappy state of affairs may account for the GSH deficiency. Supplying the necessary amino acids through supplementation, however, can bring glutathione levels in diabetics back up.

In a 2011 Baylor University study, twelve patients with type 2 diabetes were provided with amino acid supplements to encourage the production of GSH. After two weeks, they showed marked improvement in levels of oxidative stress. A few years later, another Baylor research team was also able to show that bringing up glutathione levels through supplementation reversed insulin resistance in aging mice, leading them to suggest it may do the same for humans. These studies don't show us unequivocally that glutathione can normalize a diabetic's blood sugar levels, but reducing oxidative stress in diabetics is unquestionably a step in the right direction. I'm also encouraged by some anecdotal evidence that shows glutathione can help treat diabetes.

I have a patient by the name of Vinay. He was fifty-six years old and diagnosed with prediabetes first, then later full-blown type 2 diabetes. As per the usual treatment protocol, his doctors put him on insulin. They started him on one shot a day, and it did nothing for his blood sugar levels. Then, they put him on two insulin shots a day, and, again, nothing. By the time they put him on three shots a day, his blood sugar levels were still not dropping. Other medications didn't work either. Vinay had even cut out most carbohydrates (including sugar) and started running at least five miles a day, but still no results. His weight also steadily grew, until he had gained over forty pounds, further complicating the overall issues.

During this time, he was introduced to topical glutathione. Within a few months of using it (along with a half dose of one diabetes medication), his blood sugar levels finally started to budge. Over time, the results from his hemoglobin A1C test, a diabetes measurement tool, went from 14.2 percent all the way down to 5.5 percent, the normal range. He's no longer classified as diabetic.

CANCER

Cancer, as you undoubtedly know, is the great medical puzzle of our time. Seeking a cure and an ironclad prevention strategy is our collective heart's desire. And, of course, even as we edge closer to controlling cancer, it's still an enigma, partly because, as we have learned more about the disease, it's only become clearer that it's not really one disease at all. According to the Cancer Research Institute, cancer is as many as two hundred different illnesses. And the causes can be nearly as varied, ranging from genetics, environmental exposure, and lifestyle habits to even viruses and bacteria.

One thing we know that happens in the development of cancer is that damage to a cell's DNA starts the cascade of events that lead to disease. In the rough-and-tumble world inside our body, with cells constantly growing and dividing and free radicals gunning for their electrons, cells get damaged all the time. Usually it's no problem: Those cells die off and are replaced with new healthy cells. Sometimes, though, things do not go as planned. The "bad" cells don't die off and instead divide into new cells that can proliferate without stopping, forming tumors and eventually invading nearby (or even far-off) tissue. This is how malignant tumors start.

Most cells are prevented from ever reaching this point by the anti-oxidants that subdue damage-inducing free radicals. Glutathione is among them, but glutathione also plays another role in cancer prevention: By removing carcinogens in its role as a detoxifier, it helps stop cancer before it can start. Glutathione also assists the body in defeating viruses, another way it has a protective effect.

When you consider that in 2018, there were an estimated 1,735,350 cases of cancer diagnosed with over 600,000 people dying of the disease in the United States, it's essential that we all do our best to marshal our body's cancer-preventive forces. And, as with other diseases, that includes maintaining healthy GSH levels and even enhancing your levels through supplementation. It's important to note, though, that if you have or have had cancer, it's necessary to use caution with supplementary glutathione.

When it comes to cancer, glutathione has a dual role. It protects against it, as I've just noted. But excessive glutathione can also interfere with chemo-therapy drugs. Many types of tumors have been found to have elevated levels of glutathione and it's not altogether surprising since glutathione helps cells divide—and that's what cancer cells are known for, dividing *too* prolifically. Many chemo agents attack cancer cells by generating oxidative stress and here, too, GSH can interfere by doing its normal job of stamping out free radicals. Some research has shown that depleting glutathione from cancer cells helps sensitize tumors to anticancer drugs, increasing the drugs' effectiveness.

In a 2013 journal article summarizing glutathione's role in cancer progression and prevention, the authors called GSH a "double-edged sword." On the one hand, they wrote, glutathione has the potential to be used as a therapeutic agent to help protect against cancer. On the other hand, using medical techniques to deplete glutathione has already been shown to help kill off cancer cells.

Does this mean that increasing your glutathione through natural means such as diet and supplements will put you in harm's way if you don't have cancer? "No," says Vishwanath Venketaraman, PhD, a professor of microbiology/immunology at the College of Osteopathic Medicine of the Pacific Western University of Health Sciences in Pomona, California, a researcher you'll learn more about beginning on page 54. In an interview, Dr. Venketaraman told me, "Regular intake of glutathione can help reduce oxidative stress. Any condition that is because of oxidative stress, whether it's aging or cancer, can all be minimized because of glutathione." I agree with Dr. Venketaraman on this point. Glutathione helps reduce all the carcinogens and the viruses that we know can cause cancer. What you want is something that will do all the dirty work needed to help prevent cancer in the first place, and that is glutathione.

VIRAL AND BACTERIAL INFECTIONS

As this book was going into production, the world suddenly became enmeshed in the COVID-19 pandemic. I want to be careful here because at this time we still don't know enough about the coronavirus to make any predictions about *anything*, let alone about how glutathione might fit into the picture. But let me say this: All viruses and infections attack your immune system and weaken your body and its ability to fight diseases. It is well-known that glutathione levels are low in patients with many types of active infections; what we still don't know is whether the COVID-19 virus infects people with low glutathione levels. And here is another question: If you have an intact immune system and normal glutathione levels, when exposed to a deadly virus such as COVID-19, will your body be able to fight it off and keep you safe? These are ideas I hope will be explored in the future. As for now, let me tell you what we have learned so far about glutathione as it relates to infections.

Infectious disease experts have been warning us for years that the days of effective antibiotics are waning (remember antibiotics treat bacterial not viral infections, which is why they weren't a help in the COVID-19 pandemic). And now, as medical practitioners report seeing more and more antibiotic-resistant infections among their patients, the warning has become even more urgent. There are many things we can do to help turn the tide— always finishing a doctor-prescribed course of antibiotics is one of them— but it's also important to think innovatively about other ways to prevent and treat infections. Could glutathione be one of them? Evidence suggests it can. GSH regulates immune cells, helping them fight infection, and it may even have inherent antibacterial properties, particularly at higher concentrations. In fact, GSH's structure resembles that of penicillin, and it has been called "glutacillin."

Glutathione also has a preventive effect in its role as an antioxidant. One by-product of oxidative stress is inflammation, and inflammation creates just the kind of environment that viruses and bacteria tend to favor. As in so many conditions, when glutathione levels are low, infection can creep in. We know, for instance, that cells infected with hepatitis, influenza, and HIV have decreased intracellular GSH. Low glutathione has even been associated with low survival rates in patients with HIV.

Some of our best understanding of the part glutathione plays in HIV and another intractable infectious disease, tuberculosis (TB), comes from the work of Dr. Venketaraman, whom I mentioned earlier. Dr. Venketaraman has been studying glutathione for many years and has published extensively in peer-reviewed science journals on the topic.

His investigation of glutathione in relation to TB is particularly interesting. Around the globe, tuberculosis is among the leading causes of death from infectious agents, and the leading cause from one single infectious agent—a bacterium called *Mycobacterium tuberculosis* (*M. tb.*). The World Health Organization reports that 1.3 million people a year die from the disease.

But even more people—perhaps as many as ten million—are infected with TB every year. "It's a major global burden," says Dr. Venketaraman. "Also, there are findings that not all people that get infected with *Mycobacterium tuberculosis* have an active disease. Among people who get infected only ten percent suffer from active disease. In the other ninety percent, the bacteria remain latent in the body. A third of the world population may have latent TB." In most cases, latent TB stays dormant. But if the immune system of a person who has it becomes compromised, the bacteria can activate, and the disease can erupt like a once-quiet volcano.

Among the most vulnerable to TB infection and activation of the latent form of the disease are people infected with HIV and/or have full-blown AIDS. The same is true of people with type 2 diabetes, who are two to three times more susceptible to TB than their healthy counterparts. Why these two seemingly disparate groups? It could be due to one factor that they *do* share: low levels of glutathione. In other words, both diabetics and people with HIV/AIDS are lacking one of the best-known safeguards against infection.

The lack of glutathione in HIV/AIDS patients and diabetics is likely due to two things that take place in the advent of disease. One is increased oxidative stress, triggered by cytokines, inflammatory substances sent out by the immune system to battle invaders. The other is an inability to properly "build" glutathione. "Glutathione is made of three amino acids: glutamine, cysteine, and glycine," explains Dr. Venketaraman. "But within the cell you need two enzymes to form glutathione—and those two enzyme levels are compromised in people with HIV and type 2 diabetes."

If the body's own glutathione stores are being overwhelmed by oxidative stress and its ability to make more is compromised, what will happen if you give the body an assist? Dr. Venketaraman and his team decided to find out.

They knew from previous work that GSH was essential to fighting TB. "Our research has shown that if you grow the tuberculosis bacteria with no immune cells in a laboratory setting, and you include glutathione in the growth medium, the glutathione will keep the pathogen from dividing," says Dr. Venketaraman. "Which means that it's almost like an antibiotic. It's directly toxic to the TB pathogen."

"We kept asking more questions," he continues. "There are a lot of immune cells in the body, like T cells and natural killer cells. What would happen if we increased the glutathione levels in these cell types? We got an answer from experimental findings on HIV patients that showed increasing glutathione in these cell types supports the inhibition of the TB pathogen growth within the macrophages [a type of immune cell]. So, glutathione has multiple roles. It can directly kill the bacteria and also act as an immune enhancing agent, activating the immune cells."

Dr. Venketaraman has also recently looked at the potential effect of using glutathione to treat tuberculosis in people with type 2 diabetes. His research group took cells from diabetics and treated the cultured cells with both glutathione and low levels of antibiotics. The combination completely cleared the *M. tb.* infection. "Which means glutathione can be given as added therapy," says Dr. Venketaraman. "One of the common problems with TB treatment, unlike other infections where you go to the doctor and the doctor gives you antibiotics for seven days, then you're done, is that TB treatment can last for six to nine months. And it's not one medication but three or four drugs. There are also side effects of the drugs. It's possible that if [patients] are also given glutathione treatment, they could minimize their treatment from six months and experience fewer side effects."

There might be preventive effects, too, says Dr. Venketaraman. "I believe the vulnerable population who have HIV and diabetes, if they could consume glutathione on a regular basis, it may protect them against any opportunistic infections."

You can see why I have been following Dr. Venketaraman's work so closely! His findings are very exciting, and because his work with glutathione is ongoing, there are sure to be more positive developments to come. I believe we can also expect to see some glutathione-related advancements in the area of chicken pox, herpes, and shingles. I can say from experience that GSH is effective for treating shingles and herpes. Patients of mine have seen sores (and cold sores) go away within forty-eight hours. This is particularly exciting when it comes to herpes, thought of as an "incurable" disease. It will definitely be among my next areas of research.

FEEL BETTER: GLUTATHIONE CAN HELP YOU FIGHT THE FLU

WITH ALL THE FOCUS on the coronavirus, we cannot forget about the flu, also a potential killer. In the 2019–2020 flu season, the Centers for Disease Control and Prevention estimated that 26 to 36 million people would become ill with the flu and that deaths would number up to 36,000. With the flu, prevention is key, and simply having adequate amounts of glutathione in your body's arsenal can help you fight it off by giving the immune system a big assist in doing its job. And, as it turns out, if you do catch the flu, glutathione can also help you recover more quickly. That's what Italian researchers at the University of Pavia and University of Genoa found when they tested the healing power of GSH. In their study of 262 men and women, half received N-acetyl cysteine (NAC), a precursor to glutathione (in other words, it helps the body make glutathione), and half received place-bos. Both groups took two tablets a day for six months, and none of the participants had had a flu vaccine for two years.

The researchers were already quite aware that glutathione had unequivocally reduced flu severity in mice. So, how did it work in humans? Very well. The men and women taking the NAC had a significant decrease in the frequency of flu-like episodes, illness severity, and length of time they were stuck in bed trying to get well. An unqualified success!

I inadvertently conducted a similar (though less controlled) study myself when seventeen of my patients using GSH for cosmetic antiaging purposes all came down with bad cases of this season's nasty flu virus. To my surprise, these patients either contacted me to ask whether glutathione could help their body repel the virus or just decided to use glutathione on their own to see whether it would help them feel better. When they reported back, all of them said that their fever broke within twenty-four hours, and they quickly felt substantially better and more alert. Most of them also noted that the residual effects of their stuffy nose and/or congestion lingered for a few days, but the fever, lethargy, and body aches were gone in twenty-four hours. This was a huge aha! moment for me.

And it also begs the question, Can glutathione help *prevent* the flu? I think it can and I'll tell you why. What is the best way to defend your house against thieves and intruders? Is it to buy a shotgun and camp out twenty-four hours a day waiting to shoot them once they enter your home? Or is the best defense to deter all unwanted intruders with a proper alarm and security system, and keeping your windows and doors locked at all times? Obviously, the latter of the two options is the better and safer choice.

Glutathione is a security system for your body. It's the best-known defense that your body can employ. I suggest that you let glutathione lock all of your doors and windows for you, keeping the bad guys out. Let it do its job so that in-truders have less of a chance at getting in.

.... • ....

HAPPY FEET

Have you ever heard of Hong Kong foot? Like most people, I hadn't either until a colleague was visiting Asia, met a man who was suffering, and offered him some glutathione as a remedy. A few months later, we found out how it worked.

Hong Kong foot is widely regarded as an extreme and mutated form of athlete's foot that is resistant to all drugs and keeps mutating to outsmart the medical community. Like athlete's foot, it's a contagious fungal infection that causes a scaly rash, itchiness, stinging, and burning. The condition is quite common in China—reportedly affecting 20 to 50 percent of the population. And there's a story there.

Legend has it that after the British won the first Opium War in China (1839–1842), soldiers occupied Hong Kong. Dressed as they were in military boots and given the heat and humidity of Hong Kong, many of the Brits developed itchy, fungal rashes on their feet, something never seen before in the colder, drier climes in England. The name "Hong Kong foot" took hold and now it's how this severe form of athlete's foot is commonly referred to in Asia.

Now, back to my colleague's encounter with a longtime sufferer of the dreaded infection. She was on a business trip when she was introduced to a certain Mr. Wu, a very wealthy man from China. It became apparent that Mr. Wu could not walk very far, and the reason was Hong Kong foot. The man had had it for years and tried every possible drug and therapy to make it go away. All the money in the world could not cure his feet.

Feeling sympathy for the man, my colleague offered him her bottle of glutathione, hoping it would help. A few months later she was sitting in a meeting with another business associate from China, when he received a text from Mr. Wu. Mr. Wu had taken a picture of the glutathione bottle and wanted to know if my associate could find "this girl" (my colleague) that gave him the glutathione. Why? Because within five weeks his feet were completely healed.

I know a lot about glutathione, but this phenomenal effect on infected feet was even news to me. And it's still not completely clear how it works. That said, I'm certain that glutathione helps alleviate the oxidative stress generated by the fungus, and I also believe that it is most likely doing some detox work, getting rid of microtoxins causing the condition. The results are real, and I believe we will figure out why with further investigation. In the meantime, it's a step in the right direction toward curing Hong Kong foot and other forms of athlete's foot, including the types that afflict gym-goers stateside.

········•········

AUTOIMMUNE DISEASES

Typically, the immune system is our knight in shining armor, protecting us against harm. But sometimes, in some people, the immune system overreacts and produces antibodies that cause the body to attack its own tissues. The consequence can be any number of autoimmune diseases—there are over eighty different types of them, including Graves' disease and Hashimoto's (which both affect the thyroid), inflammatory bowel disease, Crohn's disease, type 1 diabetes, psoriasis, and rheumatoid arthritis.

When an autoimmune disease strikes, it's a sign that the body is not in harmony, and the imbalance can cause your whole system to go haywire. Researchers have been puzzling over the cause of autoimmune diseases for decades. There is almost certainly a genetic component, and as experts have noted in the *New England Journal of Medicine*, "Even in a genetically predisposed person, some trigger—an environmental exposure or a change in the internal environment—is usually required [for autoimmune activity]."

Glutathione is intimately connected to the immune system. As Dr. Jimmy Gutman notes in *Glutathione: Your Body's Most Powerful Protector*, GSH is "food" for the immune system: It helps keep the immune system operating properly by enhancing the activity of certain immune cells and by fighting free radical activity within those cells. This second aspect of glutathione's protective power may be particularly important when it comes to treating autoimmune diseases with supplementary GSH. Intracellular glutathione can be important for a surplus of reasons, but mostly because it is the main antioxidant for the mitochondria. Mitochondria, as you may remember, are the tiny organelles inside each cell that convert nutrients into energy. The degenerative process of autoimmune diseases destroys the mitochondria in affected cells. Glutathione helps to thwart this destruction and counter the oxidative stress known to exacerbate autoimmune disorders.

There have been some interesting developments in how glutathione helps ease certain autoimmune disorders that are worth talking about in depth. Here are two (I'll talk about a third autoimmune disease, psoriasis, in Chapter 7).

Hashimoto's Disease

It may be small, but the tiny, hormone-producing gland that sits at the bottom of your neck—the thyroid gland—can lead to big problems when diseased. Hashimoto's occurs when your immune system attacks your thyroid, a critical gland that helps regulate metabolism, heart rate, digestion, and several other processes in the body. When under attack, the thyroid slows down (hypothyroidism) with consequences you can both see and feel: fatigue, muscle aches and stiffness, dry skin, weight gain, even depression and memory lapses. Like many autoimmune diseases, Hashimoto's runs in families, though it's also been linked to radiation exposure and excessive iodine intake.

And glutathione? Hashimoto's is indeed associated with lower levels of GSH. This was discovered by a research team at the Iran University of Medical Sciences. The team compared glutathione levels of forty-four women with Hashimoto's disease to those of fifty-eight women with fully functioning thyroids and found that the Hashimoto's group had significantly lower—62 percent lower—levels of GSH. This is a familiar refrain in disease, and a sign that the afflicted patients were experiencing considerable oxidative stress. But the researchers also pointed out that glutathione depletion affects the signaling of certain immune cells and that low glutathione may hasten cell death, two factors that could also allow Hashimoto's to take root.

Leaky Gut Syndrome

I'll say at the outset that leaky gut syndrome is not definitively defined as an autoimmune disease, but some research suggests it's linked to other auto-immune disorders. Many of my patients who have autoimmune problems are also dealing with leaky gut syndrome, and addressing it is one of the first stops in navigating and managing their autoimmune issues.

If you have never heard of leaky gut, suffice it to say that it's just as awful as its name sounds. When the lining of the small intestine becomes overpermeable it leaks, allowing tiny undigested food particles, toxins, and pathogens to make their way into the sterile environment of the bloodstream. This can trigger systemwide inflammation, leading to food sensitivities, pain, brain degeneration, and ultimately other autoimmune diseases.

Low glutathione levels can make a person more prone to developing leaky gut syndrome. That's because, among its other protective actions, glutathione plays an important role in reducing intestinal inflammation and improving gut barrier strength. Unfortunately, your gut isn't the only barrier that can become overpermeable when your glutathione system is depleted. Some people develop a leaky respiratory tract or a leaky blood-brain barrier. With this in mind, it's vital to support your glutathione levels both as a preventive measure and, if you have been diagnosed with leaky gut, treatment for the condition. One of my patients, Janney, found that glutathione erased the fatigue and digestive problems she was experiencing from leaky gut syndrome, so I can say with some confidence that GSH has a very positive effect. (Read more about Janney's experience with glutathione on page 111.)

There is yet another connection between glutathione and autoimmune diseases. In an inspiring TEDMED Talk, Mark Hyman, MD, whom you may know as the author of *The Blood Sugar Solution* and *Food: What the Heck Should I Eat?* and other blockbuster books, talks about how he went from healthy and thriving one day, a hundred-mile-a-day cyclist, to barely able to walk up a flight of stairs. He had digestion problems, rashes, sleeplessness, brain fog, weakness—you name it.

Dr. Hyman is a practitioner of functional medicine, which means that he approaches health by looking at the imbalances in the network of someone's biology, rather than simply diagnosing a disease based on generalized signs and symptoms. In other words, he bases treatment on the system not the symptoms. But at the time, he turned to traditional medicine to find out what was wrong with him, making the rounds of an array of specialists. He found no answers, but was prescribed a lot of different medications in the meantime.

When Dr. Hyman turned inward to look at the various systems in his body, he began to see things more clearly. And one thing that he saw was that his glutathione levels were low. Another thing he saw was that his mercury levels were high. Many years earlier, Dr. Hyman had spent considerable time practicing medicine in China—and breathing in the country's notoriously polluted air. He had worked at a hospital in an inner-city neighborhood, too, where the environment was also probably fairly polluted. Now years later, he tells his TEDMED audience, the heavy metals in his system—along with some other systemic problems—were taking a toll, creating many symptoms compatible with an autoimmune disease.

One theory of the development of autoimmune disorders proposes that tissue damaged in one way or another triggers an attack of the immune system. In Dr. Hyman's case, mercury damage may have been the culprit and his low glutathione status the complicating factor: There wasn't enough GSH to go around in the case of extensive heavy metal exposure. Dr. Hyman's experience as both a physician and a patient led him to be one of the earliest and most avid champions of glutathione. As far back as 2010, he wrote in the *Huffington Post*, "Glutathione is the most important molecule you need to stay healthy and prevent disease . . . There are now more than 89,000 medical articles about it [a figure that is much higher now]—but your doctor doesn't know how to address the epidemic deficiency of this critical life-giving molecule."

OBESITY

There are many reasons we gain weight as we age—loss of calorie-hungry muscle mass is one of them—but the downturn in glutathione production as we grow older may also contribute to newfound pounds. Glutathione is instrumental in ensuring that fat gets burned by the mitochondria, those tiny energy factors inside cells. Less glutathione, less fat burning.

That, at least, was the conclusion of a combined animal and human study conducted by researchers at Baylor University and published in 2013. The scientists looked at glutathione and fat burning from three different angles. The first group of subjects were young mice, who had been given a compound that inhibits the synthesis of glutathione. With their GSH forces depleted, the young mice had a decreased ability to burn fat. The second group of subjects were older mice. These older mice didn't need to have their glutathione artificially depleted because, due to aging, they already had lower levels of the antioxidant. So, what the researchers did instead was to enhance the older animals' ability to make GSH by feeding them gluta-thione building blocks. Not surprisingly, as their glutathione increased, so did their fat burning. And they not only lost weight, they decreased their risk factors for diabetes.

Animal studies can give us insight into what might provide health ben-efits for humans, but the best way to find out what works in humans is to test . . . humans, of course. That's why the most interesting aspect of the Baylor study was the testing they did on eight elderly people. The group were given cysteine and glycine supplements for two weeks to correct their age-related glutathione deficiency. As a result, the elderly eight returned to burning fat just as efficiently as eight young control subjects.

The outcome of the Baylor study is encouraging, but I want to stress that glutathione-raising supplementation is only one part of the weight-control story here. When you consciously eat to elevate your GSH levels, you are most likely going to slim down. We all need a full arsenal of gluta-thione to keep the home fires burning, and eating a glutathione-smart diet can help with that. But, perhaps even more important, this kind of approach to eating, as you will learn when you dive into Action Plan One (page 133), minimizes your intake of the empty, excessive calories that cause weight gain in the first place. In their place are nourishing, nutrient-rich calories (such as those found in healthy fats and low-carb vegetables) that facilitate weight loss. By making glutathione your focus, you'll fight obesity and all the ills that accompany it.

LIVER DISEASE

Many diseases affect the liver, that clearinghouse for toxins and wastes we so greatly depend on. Some of them, such as hepatitis (A, B, and C), are caused by viruses. Some, like cirrhosis and fatty liver disease, can be a by-product of drug and/or alcohol use. There are also inherited conditions, such as Wilson's disease, that prohibit the body from excreting copper. In Chapter 4, you'll learn about how integral glutathione is to the liver's job of detoxifying the body. But since we're on the topic of disease here, let's talk about how GSH can also help ease certain liver disorders. Case in point: hepatitis B.

Hepatitis is a condition characterized by liver inflammation and usually caused by a virus (although heavy alcohol use, certain medications, and toxins can also trigger hepatitis). When it goes untreated, hepatitis can lead to liver cancer or liver failure. There are three types of viral hepatitis: A, B, and C. Vaccines are available for hepatitis A and B, but the disease still has a foothold in the US, especially types B and C. The Centers for Disease Control and Prevention estimates over twenty-two thousand B infections and over forty-four thousand C infections each year. Both can be contracted primarily through shared body fluids. (Hepatitis A, about seven thousand cases a year, is more likely to be contracted through food or drink contaminated with feces.)

Based on the knowledge that glutathione plays an important role in repairing injury to the liver, medical researchers in China decided to see if it would have an effect on patients with hepatitis B. They divided 104 patients into two groups and treated one group with intravenous glutathione once a day for eight weeks. The other group received a placebo drip. When all the data were tallied, it was clear that glutathione had helped reduce inflammation and the activation of messenger chemicals that impair liver function.

Glutathione has also been shown to improve the symptoms of a condition called nonalcoholic fatty liver disease. NAFLD occurs in people who drink little or no alcohol yet suffer liver dysfunction (there is also a related disorder known as alcoholic fatty liver disease). Specifically, as the name implies, their livers store too much fat. The disease often goes hand in hand with diabetes, obesity, as well as high cholesterol and high triglycerides, both risk factors for heart disease. In a small percentage of cases, NAFLD progresses to cirrhosis or liver cancer. But even in its mildest form, the disease can make you feel tired and uncomfortable. The cause of NAFLD is unknown, but it affects about 25 percent of the population.

My particular interest in fatty liver disease stems from a study that used topical glutathione supplied by our lab. There have been thousands of published, peer-reviewed studies on glutathione, all of which have added greatly to our knowledge of the antioxidant. What has helped us understand the power of this mighty molecule, too, are stories from individuals recounting their own experiences as well as small studies conducted by boots-on-the-ground clinicians. These more anecdotal types of evidence—including the research I'm about to cover here—are very valuable in helping us learn about the real-world uses of glutathione.

Ray Solano, PharmD, is a clinical pharmacist and board-certified nutritionist who practices in Cedar Park, Texas, where he runs a specialty compounding pharmacy. The pharmacy also has a clinic where Dr. Solano and his colleague Jim Meyer see patients with assorted health issues. Dr. Solano's pharmacy has been formulating IV glutathione for years, mainly for the use of children with autism (see page 32 for more on glutathione and autism), and for people exposed to heavy metals and other toxins. Dr. Solano himself has a personal interest in the antioxidant. "I've had genetics snips that tell me I'm glutathione deficient," he says; he makes up for the deficiency with supplementary GSH. "I can see the difference in the reduction of toxins, histamines, and allergens." And he adds jokingly, "I've been a glutathione junkie!"

Recently, our lab supplied Dr. Solano and his colleagues with our topical form of glutathione so they could study its effects on people with nonalcoholic fatty liver disease. When his clinic decided to test glutathione in patients with NAFLD, they weren't shooting in the dark. Research has shown a connection between low glutathione and NAFLD and some scientists have had success using glutathione to positively affect factors that promote the progression of fatty liver disease. For instance, a Japanese study published in 2017 found that a marker of the disease—an enzyme called alanine aminotransferase (ALT)—significantly improved in thirty-two patients who received a specialized oral glutathione treatment daily for four months. The drop in levels of ALT indicated improvement in the buildup of fat cells that is the hallmark of NAFLD.

Dr. Solano's group found glutathione improved patients too. The study involved five people with NAFLD who followed a three-month regimen that combined treatment with transdermal glutathione and a low-carbohydrate ketogenic diet (see pages 137–138 for more on the keto diet). The main focus of the study was triglycerides, which are typically high in people with fatty liver disease. Triglycerides are a type of fat in the blood we use for energy, but when elevated are a risk factor for heart disease and stroke.

When the NAFLD patients were tested at the end of the three-month period, they not only had significantly lower levels of triglycerides, they felt notably better, and had lost a considerable number of pounds. Would the glutathione have worked without the diet or vice versa? Or was it a combination of the two that proved so beneficial? Most likely the combination, says Dr. Solano.

For one thing, ketogenic diets have been shown to increase glutathione in the mitochondria (see page 137) so GSH was getting a boost in two different ways. Also, oxidative stress, if you may remember from earlier in this chapter, has an injurious effect on lipids, so by reducing oxidative stress, GSH can lower triglycerides—as can a keto diet. The keto diet was likely the biggest factor in the study subjects' weight loss. But in Dr. Solano's estimation, they may not have been able to stay on the diet if it weren't for the glutathione. "We have found that triglycerides usually stay stubbornly the same in patients because it's very difficult for patients to stay on any diet. They were either noncompliant, or those who were compliant would give up after a while," he says. One reason may be mood. "People who have high levels of triglycerides tend to be a little depressed, so they don't always have the strength to be able to change their behavior."

Glutathione can help with that. Some research has shown that low glutathione levels are linked to depression; therefore, raising levels may improve mood (see page 30 for more on GSH and depression). "Glutathione pathways are a critical part of the biochemistry and they affect serotonin levels and dopamine levels," explains Dr. Solano, referring to the neurotransmitters that help regulate mood. That's most likely why, with glutathione added into the mix, the NAFLD patients had an easier time sticking to their diet, hitting the high-triglyceride problem from two angles and resulting in a good outcome.

It's also quite possible that the combination worked so well because a ketogenic diet creates a state in the body, called ketosis, where it burns stored fats. During this process, the body up-regulates the synthesis of glutathione, meaning you have more of it. That may be another reason the GSH-keto diet combination was a winner for Dr. Solano's patients.

To be honest, I'm not as concerned about what had the most impact on Dr. Solano's subjects, the keto diet or the glutathione. I'm more concerned with the end result. And their results were shocking. The keto diet has been effective for a significant segment of the population, and the research on using it to treat seizures, especially in children, is quite intriguing.

Dr. Solano's study augurs well for anyone who has fatty liver disease and/or high triglycerides, including people with heart disease and diabetes, which are also often linked to high triglyceride levels. Now we know there may be a way to reduce them with a combo of keto and glutathione.

MACULAR DEGENERATION AND CATARACTS

Most of the popular reporting on free radicals makes the connection between the rogue molecules and two predominant ills: cancer and aging of the skin. But as I've noted throughout this chapter, oxidative stress does indiscriminate damage; therefore, it's linked to *many* kinds of conditions. One of them is macular degeneration, a primary cause of vision loss in people over age sixty.

The portion of the eye that degenerates in macular degeneration is the retina, light-sensing tissue in the back of the eyes. Deposits or abnormal growth of blood vessels in the retina eventually lead to diminished or distorted sight. Macular degeneration can have a hereditary component, but being a smoker or having high blood pressure, high cholesterol, or diabetes can also make you more susceptible to this debilitating eye disease. So can being white, female, and blue-eyed. Lack of glutathione plays a role, too: Studies have shown that people with macular degeneration also have low GSH levels.

The eyes are particularly predisposed to free radical production. The retina consumes more oxygen than any other tissue in the body, which means that it's also a little free radical factory (remember free radicals are formed when oxygen is consumed for energy). The eyes are also subject to a lot of light, another cause of free radical development, and their photoreceptor cells contain plenty of fatty acids, which are easily oxidized. When your body is rich in glutathione, creating a healthy antioxidant-to-oxidant ratio, the eyes come out of all of this relatively unscathed. But as you get older and your glutathione levels drop, eye damage becomes a real possibility.

Cataracts, a clouding of the eye's lens, is another condition caused by oxidative stress. When glutathione and other antioxidants are in short supply, free radicals not only damage the proteins that make up the lens, they also make it difficult for the cells to deploy their usual repair mechanisms. Over time, the lens clouds up even more until it interferes with sight.

Several antioxidant eyedrops containing glutathione are on the market, some of which have been shown to improve vision and stabilize cataracts. Additionally, studies in mice have shown that treatment with compounds that increase GSH levels reduce oxidative stress in retinal and photoreceptor cells, making these glutathione-boosters a potential treatment for macular degeneration. Foods that increase levels of glutathione and other antioxidants can help too. I'll talk more about that in Action Plan One.

4 LAYING WASTE TO TOXINS

TROUBLE SLEEPING. RAPID WEIGHT GAIN. RAPID HEART RATE. GLUCOSE high. Energy rock bottom. Heavy menstrual periods. Overstimulation. Anemia. Brain fog. Anxiety. Feelings of despair.

Bryanna is a woman who, at the young age of thirty-two, could lay claim to this long list of health problems. The symptoms had been building for years, and although she saw several doctors through the course of her odyssey, Bryanna could not find any answers. What was wrong with her?

Perhaps it was early menopause. The symptoms seemed to fit, but when Bryanna went to a gynecologist, all the tests were negative. That was a relief, but it put her back on the merry-go-round with no diagnosis. And her problems were not letting up. Bryanna gained thirty-six pounds over the course of three years. Even when she tried working out every day, subsisting on vegetables, experimenting with different supplements and gluten-free living, the pounds wouldn't budge. In all that time, she'd barely slept. Just a few years earlier, she had been an energetic young woman thriving in her career as an aesthetician. Now she was at wits' end.

One of the people Bryanna talked to about her symptoms was Dan Holtz, cofounder of Beverly Hills Rejuvenation Center. If it sounds like the kind of place where the rich and famous go for skincare and wellness treatments, that's because it is, and Dan is well-known for his insight into how to fight aging and achieve optimal health. But even Dan was stumped by what was happening to Bryanna. He did, though, think it might have something to do with toxins. Bryanna was skeptical. "Whenever he mentioned it, it never struck a chord," she says.

One day, a lightbulb went on in Dan's brain and he gave Bryanna a bottle of glutathione. As a colleague of mine, he knew about glutathione and knew that it plays an integral role in helping the body rid itself of toxins. There seemed to be nothing to lose by having Bryanna try it, and if it worked, Dan would know that his theory about Bryanna's condition was correct. She was game.

Bryanna began by spraying her abdomen four times with transdermal glutathione. What happened next astounded her—and Dan. "Fifteen minutes later, I was knocked out and into a deep sleep," she remembers. The next morning, she woke up and thought, "What was that?" before giving herself a second dose. She slept again. And kept on sleeping on and off through the next week, getting more than the eight hours she had pined for just days before. Sleeping and eating. Sleeping and eating. When the week was up, she had, to everyone's astonishment, lost fifteen pounds.

And that wasn't all. Her brain fog had lifted, and she felt well rested and energetic. All the symptoms that had plagued her for years went away. It got Bryanna thinking. She had been skeptical when Dan suggested toxins might be at the root of her health woes, but now Bryanna started looking more carefully at her life to see where she may have been exposed to heavy metals or chemicals. The answer was staring back at her in the mirror. "I dyed my hair a lot," she says. "And I really went overboard. If the box said to wait a certain number of weeks before you dye it again, I would never wait that long. If it said leave on for forty-five minutes, I'd leave it on longer." Without even knowing it, Bryanna had been dosing herself with toxins for years.

I generally get two different reactions when I bring up the word *toxins*. On the one hand, there are people who think the idea that we're living with toxins in our system is ridiculous. "Wouldn't we be dead or deathly ill if there were poisons in our system?" they ask. The other response I get is a knowing nod. The people in this camp are constantly worried about toxins in the body, and frequently participate in detox diets and cleanses.

The truth lies somewhere in the middle. Toxins in the body are indeed real. Some of them are even produced naturally, a by-product of the various physiological operations that keep us alive and well. But we also have a very efficient system for getting rid of toxins, thanks to our liver with a major assist from glutathione. That's one reason most of us stay healthy and go on to lead long lives.

And yet it must be noted that we are living in a world where environmental toxins are rampant, and at times this is more than our body can handle. Researchers calculate that by the time we have left the house in the morning, climbed into our cars, and headed off for work, we have already been assaulted by about 129 different types of toxins. Some evidence suggests we are also being affected by toxins while we sleep. Do you sleep with your cell phone next to your bed? Radio-frequency fields of cellular mobile phones cause oxidative stress. (Might I suggest you charge your cell phone far away from where you sleep? Maybe go back to using a traditional alarm clock.) Avoiding contact with toxins in our modern-day world is impossible!

In Bryanna's case, where the exposure to toxins was unusually high (and there are thousands of stories out there just like hers), the consequences can be quite dire. Most of us, especially if we're taking care to avoid chemicals and other such toxins as heavy metals, won't need treatment with an infusion of glutathione as Bryanna did. However, even the most conscientious among us can't avoid toxins completely and that makes boosting our natural glutathione levels more important than ever.

MY WAY OF LOOKING AT MEDICINE

MOST PRACTITIONERS IN THE traditional medical field treat symptoms and ignore the system. As tempting as it is to be a symptom chaser—that is, you identify the symptoms, then treat them—I prefer a form of medicine that looks at the bigger picture.

Before graduating from pharmacy school at the University of Southern California, I studied mechanical engineering, so my brain is wired to look at how things intrinsically work. That's why I believe symptoms should be viewed as a clinical road map: They lead us on a journey to find the system in the body that is breaking down or out of balance. Symptoms in essence are like bread crumbs. Follow their trail backward and you'll find the root cause of the problem.

Medical practitioners who think this way are typically the type that treat the whole person rather than focusing on one particular problem. Those who are symptom chasers are the biggest contributors to what I see as the overprescribing of drugs in America. That might seem like an odd complaint for a pharmacist, but when you consider that 44 percent of all Americans are on at least one prescription drug and 17 percent take three or more, you can see that the numbers are remarkably (and to my mind, uncomfortably) high. In fact, many issues can be solved with natural endogenous substances—like glutathione—that help address systemic causes of health woes.

For all these reasons, I particularly appreciate the practice of functional medicine, which focuses on connecting the dots to get to the origin of medical problems. If you've been struggling with your medical treatment or are looking to start fresh with a new practitioner, I suggest looking into functional medicine. Check out the American Functional Medicine Association's website (afmassociation.com) to learn more about it.

........•........

OUR INCREASINGLY CHEMICAL WORLD

These days, there is ample reason to be on the defensive when it comes to your health. According to the Natural Resources Defense Council (NRDC), there are more than eighty thousand toxic chemicals in use in the United States, and they are absolutely everywhere: in furniture, rugs, cookware, cleaning products, makeup, shampoo, lotions, perfume, air fresheners, clothes, toys, food. You name it. Shockingly, the majority of those eighty thousand chemicals have not been tested for safety, and subsequently are not proven safe. If you smoke or live with someone who does, your exposure to toxins is especially intense—cigarette smoke has hundreds of chemicals and free radicals.

In 2017, the Environmental Defense Fund (EDF) did an interesting experiment that drives home just how fraught the world around us is. The EDF, a nonprofit staffed with scientists, policy experts, and economists, equipped ten people with specially devised wristbands that detect chemicals. The ten people were from all walks of life—there was a firefighter, a pharmacy student, a teacher, a retiree, and an airport worker among them—and they all wore a wristband as they went about their daily activities. The results were alarming. Through contact with cosmetics, personal care products, adhesives, cleaners, and other everyday items, the wristband wearers were exposed to twenty-six potentially harmful chemicals, including a highly toxic pesticide that was banned over thirty years ago.

Experts are also finding that our air and water are contaminated at levels that have the capacity to make us sick. The NRDC, for instance, has reported that the drinking water of six million Americans is contaminated with dangerous chemicals, and with environmental protection regulations being gutted in Washington, things may get worse. For the last twenty years, the American Lung Association has been issuing a "State of the Air" report, a measure of ozone and particle pollution. In 2019, the organization warned that "ozone and short-term particle pollution worsened in many cities in 2015–2017, compared to 2014–2016. Even levels of year-round particle pollution increased in some cities."

Some of the chemicals of particular concern are known as persistent organic pollutants, or POPs. These industrial chemicals include dioxins and polychlorinated biphenyls (PCBs), which although they have been banned for a while, stubbornly persist (hence the name *persistent* organic pollutants) in the environment. PCBs can still be found in the food chain, particularly in seafood; likewise, dioxins, which show up in meat, chicken, and fish. It's difficult to avoid any contact with POPs, but making wise dietary choices can help (see page 148).

Eight Toxins to Watch Out For

Short of living in a bubble, there is no way to avoid every man-made (or natural) toxin. What's more, many harmful chemicals that are present in our air and water, mostly by-products of manufacturing, are particularly hard to avoid. Sadly, I could fill hundreds of pages with the names of toxins, but that would be unproductive and just make your head spin with worry! That's no way to live.

So instead, I'm going to focus on what you *can* do, which is to be an alert consumer and avoid products that contain substances we know are not good for us. Sometimes that means reading labels; sometimes it means avoiding certain types of products altogether; and sometimes it means asking questions of the company you're buying from. Here are eight toxins that you're likely to run into and are relatively easy to identify, if only by association. That is, you may not always see them on the label, but if you know what products typically carry them, you'll have an easier time avoiding them. Remember, too, that increasing your glutathione levels is going to help you combat any toxins that, despite your best efforts, you do come in contact with.

NAME	FOUND IN	HAZARDS
Bisphenol-A (BPA)	Household plastics, food packaging (e.g., inside cans, although many are now BPA-free), water bottles, receipt paper. It's difficult to tell whether a product contains BPAs.	Linked to hormone disruption, cancer, and diabetes
Phthalates	Plastics, solvents, and personal care products, such as shampoos, nail polish, and body lotions. Look for them on ingredient lists (you'll see them in a compound name; e.g., butyl benzyl phthalate); avoid PVC (vinyl) products and plastics with a "3" in a triangle on the bottom.	Implicated in early onset of puberty and interfering with the reproductive systems of children and adults

Parabens	Cosmetics and body care products, from moisturizers and face and skin cleaners to sunscreens, deodorants, shaving gels, toothpastes, makeup, shampoos, and conditioners	Can cause hormone disruption
Flame retardants (including polybrominated diphenyl ethers [PBDEs])	Furniture, carpets, curtains, mattresses, electronics, car upholstery. Many manufacturers now label products indicating whether they have been treated with flame retardants (labeling is a law in California).	Endocrine and thyroid disruption, harm to the immune and reproductive systems, increased cancer risk, and adverse effects on fetal and child development
Perfluoroalkyl and polyfluoroalkyl substances (PFAS)/ perfluorooctanoic acid (PFOA)	Nonstick cookware, stain-resistant clothing, firefighting foam	May interfere with the body's natural hormones, increase cholesterol levels, affect the immune system, and increase the risk of some cancers
Formaldehyde	Pressed wood products, some fuel-burning stoves, tobacco smoke, lotions, shampoo, conditioner, shower gel, and some fingernail polishes, keratin hair treatments. Most exposure is through inhalation.	Increased risk of cancer; irritation to eyes, nose, throat; asthma
Azodicarbonamide	Yoga mats, flip-flops, faux leather (it helps soften material), some white flours	Onset of asthma and allergies; hormone disruption
Perchloroethylene (PCE, perc)	Dry-cleaned clothes, shoe polish, spot remover, paint products	Changes in mood, memory, and attention; increased cancer risk

LIMITING YOUR EXPOSURE TO CHEMICALS

IN THE "EIGHT TOXINS to Watch Out For" chart, I name specific types of products often made with harmful chemicals. Here I want to add a general list of products that can be the source of toxic exposure. It's not that you shouldn't purchase any of these products (though some of them, such as plastic water bottles, you may want to skip); rather, it's just a reminder to be vigilant when you do buy them. Look for the most natural and, whenever possible, certified-safe versions you can find.

Cleaning products
Insect repellants
Plastic food containers
Plastic water bottles
Nonstick cookware
Carpets
Perfumes
Air fresheners
Antiperspirants
Laundry detergents and softeners
Dry-cleaned clothes
Mattresses
Beauty products
PVC toys
PVC shower curtains

HEAVY METALS (NOT THE ROCK 'N' ROLL KIND)

Chemicals are not the only contaminants that have become pervasive over the years. Heavy metals are all around us too. Heavy metals are dense, naturally occurring metals, such as arsenic, cadmium, chromium, lead, and mercury. They don't biodegrade and can be toxic when they collect in body tissues. These metals can also generate free radicals, causing oxidative stress and its subsequent aftermath, cellular damage. Heavy metals, in fact, can be so damaging that they're associated with cardiovascular disease, neurologic and neurobehavioral disorders, diabetes, hearing loss, blood disorders, and various types of cancer. And, of course, as well-documented cases of lead poisoning in children have shown, they can cause developmental delays too.

So, how do heavy metals get into our body? In many different ways. Sometimes metals from industrial production leach into the air we breathe, the soil that produces our food, and the water that comes out of our taps. And industrial production is pretty pervasive — it includes everything from refineries, coal burning in power plants, petroleum combustion, and nuclear power stations and high-tension lines to plastics, textiles, microelectronics, wood preservation, and paper-processing plants. But although industry can shoulder most of the blame, heavy metals in our environment can also come via natural means, distributed through wind and rain and even volcanic eruptions.

Once these toxins are in the atmosphere, we can unknowingly inhale them and swallow them. Heavy metals, particularly mercury, are common contaminants in fish; the fattier the fish, the more likely it is to have metals (see page 148 for more on avoiding heavy metals in fish). In 2019, *Consumer Reports* tested forty-five popular fruit juices on the market and found that twenty-one of them had one or more heavy metals at a level of concern. Heavy metals have been found in herbs and medications too. So, even seemingly innocuous foods, drinks, and supplements can pose an unexpected hazard.

WHEN FIREFIGHTERS NEED RESCUE

IT'S OBVIOUS THAT FIREFIGHTERS put their lives on the line every time they're called into action. But the dangers go well beyond the risk of being trapped by flames. Firefighters are exposed to an inordinate number of contaminants when they're on the job, through both inhalation and skin contact. Chemical absorption goes up 400 percent for every five-degree increase in temperature, exponentially increasing the risk to a firefighter's health. Even handling equipment back at the station can be a hazard, and firefighters often bring the carcinogens they're exposed to in the field back to their homes and cars, prolonging contact.

All this exposure takes a toll. In 2015, the National Institute for Occupational Safety and Health (NIOSH) completed a multiyear study of nearly thirty thousand firefighters from the Chicago, Philadelphia, and San Francisco fire departments. The study found that firefighters had higher rates of cancer (9 percent higher), particularly digestive, oral, respiratory, and urinary cancers, than the general population. They were 14 percent more likely to die from cancer. Firefighters were also at 100 percent increased risk of developing malignant mesothelioma, a rare type of cancer caused by exposure to asbestos. Two other disturbing findings: the chance of a firefighter being diagnosed with lung cancer or dying of the disease increased with the amount of time spent at fires, as did the chance of dying from leukemia.

When I learned about what was going on with firefighters, I was horrified that our first responders were even more at risk than most of us know. And I couldn't help but wonder whether glutathione could help. It's well known that, as the body's prime detoxifier, the antioxidant helps clear contaminants from the system. Would dosing with GSH make a difference in firefighters' health? We decided to find out.

The medical staff at two firehouses in Southern California has agreed to let us test the effects of using glutathione solution on firefighters. The accumulation of carcinogens (products of combustion) and other toxins has long been a concern among those who work with or work as first responders. As Kevin Garling, a retired district chief of Puget Sound Regional Fire Authority and a career firefighter for thirty-three years, told me, "Due to our crucial concerns, I've been involved in health protocol discussions on how to fight the disease epidemic among firefighters, and implementing new 'in the field' procedures to lessen the effect of carcinogens." Kevin's department also partnered with the University of Washington in a research study to gather baseline data on exposure risks, and he stated, "Although fires move fast, any solutions are slow-moving, and we are desperate for answers."

For our study, we will be selecting twenty firefighters from each station. The firefighters will use glutathione solution for three months, during which time they'll undergo regular blood tests as well as tests to measure their liver enzymes (a marker of liver damage). We'll be able to tell whether added glutathione helps them excrete more toxins and, by having them fill out a questionnaire, also be able to see whether it affects their anxiety and energy levels.

Although it will be a while before the results of this study are in, we have some early indication that glutathione can help firefighters. Mark Jones, a deputy chief in Washington State, has been a case study and his physician was thrilled to see how much his glutathione levels rose and how drastically his liver enzymes came down with regular GSH treatment. Previous attempts to lower his liver enzymes were unsuccessful, but GSH showed results rather quickly.

What do liver enzymes have to do with good health? When your liver is damaged and/or overworked, your body will produce more liver enzymes to help the organ keep up with its assigned duties. So, consistently high levels of liver enzymes are generally a sign that your liver is in trouble and that there may be toxins slipping through the cracks, not getting detoxified, and circulating in the body where they can do damage, including increasing cancer risk. The only way liver enzymes get back to normal is to reduce toxic load, which is what glutathione is so adept at doing.

Mark Jones's good test results augur well for the ability of GSH to help firefighters fight one of the occupational hazards of their profession. Like all grateful citizens, we want to see these first responders live long and healthy lives.

.... • ....

THE BODY'S "HOMEMADE" TOXINS

Any processing plant is going to have waste products and that includes our own body, which, when you think about it, is the busiest factory around. It's always making or transforming something, be it food into energy or amino acids into proteins or proteins into muscle tissue. Some of the by-products of these processes can actually be quite toxic and need to be spirited out of the body, just like chemicals and heavy metals.

For example, women produce estrogen every single day. The hormone not only regulates the reproductive system, it also plays a role in the health of the heart, breasts, skin, bones, and brain. As it goes about doing its work in the body, estrogen gets broken down into different substances called metabolites. Your body can break down estrogen into 2- and 4-hydroxy-estrone metabolites, which are benign. Or it can transform estrogen into something called 16α-hydroxyestrone, which is actually quite toxic and known to increase the risk of cancer. The 16α-hydroxyestrone is easy for the body to make, but it has to do extra work to make the nontoxic 2- and 4-hydroxyestrone metabolites.

Thousands of reactions take place where the body actually has to choose which pathway to use and therefore which by-products it will produce. A healthy diet is one way to help "sway" those decisions. For instance, one reason health experts have long been promoting cruciferous vegetables, such as cauliflower, broccoli, and Brussels sprouts, is that they supply the nutrients women need to metabolize estrogen into nontoxic metabolites. This is the preferable option. But when that option doesn't exist and estrogen gets broken down into 16α-hydroxyestrone instead, your body must be able to get rid of it. That's where how much glutathione you have on hand can really make a difference.

GUARDING AGAINST
MEDICATION-INDUCED TOXICITY

I HAVE AN UNCLE who works as a chef in Napa Valley. When he came to visit me not long ago, I was alarmed. His weight had ballooned up to 220 pounds, he had become diabetic, and he was taking four different medications to treat it. In the realm of fasting blood sugar levels, if you score 5.6 to 6.9 percent mmol/L on an A1C test, it's considered prediabetes. If you have 7 percent or higher, it's diabetes. My uncle's was *141*, so things were pretty bad.

Given the state of affairs, he was also taking quite a bit of insulin. Insulin is actually the most toxic thing you can put in your body. You can't survive without insulin, but too much of it can kill you. You must be very careful with insulin. Its job is to push glucose, the sugar you need to produce energy, into your cells. Yet if there's too much of it, the body says, *Wait a second, you're giving me too many signals! I'm not going to listen to you anymore.* That's when the body becomes insulin resistant and many problems follow.

Alarmed by my uncle's condition, I gave him some glutathione, hoping that the antioxidant would help his body cope with the oxidative stress created by the insulin he was taking. When I saw him four months later, I was shocked. He had lost thirty pounds, was no longer taking insulin, and had dropped down to two medications. His A1C test score dropped to 5.5. As a chef, it's been hard for him to give up wine and he still drinks it every day, but his body seems to be able to handle it.

Here's what I think made the difference. My uncle was trying to change his diet and exercise, but it was not until he started using glutathione and saw some minor success that he got motivated to eat better and work out on a regular basis. It was the combination of these three things that likely did the trick for him; diet and exercise alone cannot change an A1C score from 14.1 to 5.5 in four months.

The insulin my uncle was taking was causing too much oxidative stress in his body. And it creates a vicious cycle. The more oxidative stress in your body, the more resistant you become to insulin. The more resistant you become, the more insulin you need to take. The more insulin you need to take, the more oxidative stress created, and so on. To break the cycle, my uncle needed to reduce the oxidative stress in his body, and he did that with glutathione.

Medication saves lives. I'm grateful for all the medications we have available to us! But medication can also have a certain amount of toxicity that some of you may not even be aware of. One of the jobs that glutathione does so well is help the body remove some of these toxic compounds.

If you're caught up in a medication spiral as my uncle was, you may need supplemental glutathione to help you get to a healthier place. Most of us, though, don't need extra GSH to help our body metabolize drugs. If you're following the healthy tips in this book, you should have enough glutathione on hand to get all the benefits and none of the toxicity from any medication you need to take.

········•········

THE CLEANUP CREW

How often do you think about your liver? If the answer is "practically never," it wouldn't surprise me. It may be the second-largest organ in the body (the skin is the largest), but the liver doesn't get the love it deserves—especially considering all the life-perpetuating functions it performs. The liver, sitting under your rib cage, plays a supporting role in making blood cells and enabling blood to clot. Through the manufacture of bile, it enables the breakdown of fats, carbs, and proteins, and storage of vitamins and minerals. It produces plenty of important enzymes too. Most important, the liver is the body's filter, akin to the little contraption at the end of the pool that collects all the detritus so that life proceeds swimmingly. (For the record, the liver isn't the only filtering organ you have: The skin is also a detox pathway, via perspiring, and the lungs, via breathing, also help you get rid of toxins.)

Blood flows through the liver, carrying with it toxins and waste products, toxins drawn in from outside the body—known collectively as xenobiotics—and waste products the body itself makes. This is a little bit like the garbage truck arriving at the sanitation facility. The truck has rubbish; the sanitation facility knows what to do with it. In this case, the liver is going to metabolize the junk; that is, break it down and get rid of it.

Metabolism in the liver takes place in phases, activated by the presence of elements both healthy and harmful—everything from pesticides to vitamins and minerals to hormonal waste products snap the liver into action. In phase 1, the liver goes to work breaking apart all this harmful matter into small pieces. During the process, free radicals are generated, but glutathione steps in and renders them harmless. That, though, is just one of GSH's jobs in the liver.

During phase 2, the liver "conjugates" the foreign particles it has just broken down, meaning that it combines those bits and pieces with other substances to deactivate any biological activity and prepare them for excretion from the body. If the conjugated toxins are water soluble, they are shuttled to the kidneys and excreted in the urine; if they're fat soluble, they'll get dumped into the intestines and exit the body through feces. Glutathione is one of the conjugating substances that binds to toxins, neutralizing them, then pushing them out the door. This is a critical part of the detoxification process and the reason that there's more glutathione in the liver than there is any other part of the body.

One thing it's important to note about the liver and the work it does to cleanse your body is that this is a *natural* process that operates on autopilot. It doesn't need a particular diet to make it work; it happens on its own! I am not against detox diets per se: I believe that they can help you get on a path toward healthier eating and that's incredibly valuable. But the idea that you need a particular diet or drink (particularly an overpriced one) to assist your liver in doing its job is a misperception. True, there are certain nutrients you need for the liver to operate to the best of its ability; that's why it's so important to build up your glutathione levels. (I'll discuss other nutrients that nourish the liver beginning on page 161.) But the idea that a magic potion will tell your liver to get busy expelling toxins isn't likely. Your liver is already hard at work.

OTHER AREAS WHERE GSH DETOXIFIES

GLUTATHIONE'S DETOXIFYING ASSISTANCE ALSO extends beyond the liver. Along the intestines and into the colon, glutathione essentially "defangs" carcinogens. It's a process much like that in the liver, where glutathione binds with toxins, turning them into water-soluble compounds that can be excreted without doing harm. However, when glutathione can't keep up with the demand, it can lead to trouble. Researchers in the Netherlands, for instance, have found a link between low glutathione in the colon and colon cancer. In a 2006 study, the researchers looked at colon tissue from three groups of people: those with colon cancer, those at high risk for the disease, and healthy people not considered at risk. Upon analysis, the people with cancer had the least glutathione, people at risk had intermediate levels, and the healthy controls had the most GSH. It's not a huge leap to surmise that the glutathione had a protective effect.

........●........

WHEN DETOX BREAKS DOWN

Bryanna's story gives you a sense of what happens when someone has acute toxin poisoning—way more than her natural glutathione reserves could handle. But having chemicals and heavy metals in your system doesn't always cause such extreme symptoms, though dangerous problems can arise over time.

Heavy metals are known to generate free radicals in the system and interfere with regular functions in the body. Long-term exposure can lead to muscular and neurological degeneration, culminating in symptoms that are similar to those suffered by people with Parkinson's and Alzheimer's disease. Lead and arsenic are the top two heavy metals found to affect human health, with mercury coming up right behind them. Mercury also happens to be the most toxic. It's in the atmosphere, released through factories, pesticides, and such things as paint fumes. Fish, especially fatty fish, often contain mercury, as do some cosmetics. Old metal fillings in the teeth are known to leach the heavy metal into the body too.

In a perfect world, we'd never ingest, breathe in, or absorb any toxic metals, but none of us lives in a vacuum. The biggest concern about having these toxins in your system is that, although your body may store them in places where the damage is limited, circumstances can change and cause you to reabsorb the molecules again. This time, though, they may go to places where they can do real harm. For example, let's say you have mercury in the fat cells in your stomach area. It's making you chubby (toxic metals can impair the metabolism), but it's not doing anything terrible. But then you lose weight and, as the excess fat gets burned, the mercury gets released into your bloodstream and intestines. Now it has the ability to travel to other places in the body, such as your eyes and kidneys.

You don't want these molecules to be flowing freely all over the place. What you want instead is to have enough glutathione to meet them head on, neutralize them, and see them to the door. And we know for sure that glutathione can do the job. I've seen it myself in a small study that I was involved in.

The study was designed to see whether increasing glutathione levels would also increase the elimination of heavy metals from the body. Thirty-four volunteers provided samples of their urine during a twenty-four-hour period and we tested it for lead, mercury, and arsenic to get a baseline level. The next day, we divided the group in two, then gave half of them our nanonized topical glutathione solution and half of them a placebo solution. Both were asked to apply the treatment liberally and periodically throughout the day. Their urine was then checked again at the end of the day.

This was a double-blind study, meaning that we researchers didn't know which volunteers were getting the glutathione or the placebo and, of course, neither did the volunteers. On day three, though, the study was unblinded and the volunteers switched places. Those who initially received the glutathione received a placebo and vice versa. Again, they were subject to a urinalysis at the end of the day.

The addition of glutathione made a significant difference. Seventy-five percent of people in the first glutathione group showed an increase in heavy metal output after treatment. Then, when the placebo group switched over to the real thing, they too (60 percent) had a dramatic increase in their output of heavy metals. With the help of glutathione, the body detoxes itself.

What I think we can learn from this study (and Bryanna's use of glutathione to remedy her symptoms) goes beyond the evidence that supplemental glutathione is an important detoxification treatment. It also supports the fact that glutathione is an essential component in detoxification and reinforces just how important it is to adopt all the lifestyle habits that will keep your glutathione-making machinery operating in high gear. Everything you learn in Part III of this book is going to help ensure that GSH is there when you need it.

HOW GLUTATHIONE "DRINKS UP" ALCOHOL

Alcohol isn't something we normally think of as toxic, especially if enjoyed in moderation. But alcohol has some toxic by-products so, whether you drink a glass of wine with dinner, favor a few beers on the weekends, are deep into cocktail culture, have been known to overdo it at a celebration or two—or all of the above—glutathione is your best friend.

The body has a fairly efficient process for dealing with alcohol. When you have, say, a margarita, the alcohol in the drink travels through your stomach and small intestine and is then absorbed into your bloodstream. How quickly depends on a number of factors. If you had a lot of tortilla chips and salsa before you drank your margarita, your cocktail will be absorbed more slowly than if you were drinking on an empty stomach. Chugging vs. sipping your drink will affect absorption (you can guess which one has a slower absorption rate), as will the concentration of alcohol in your glass. Distilled spirits like tequila slip into the bloodstream faster than wine or beer. What you mix in counts too. Alcohol combined with something carbonated, such as cola, is absorbed more quickly than drinks mixed with water.

Once alcohol is in your bloodstream, it's shuttled to the liver, where it's met by an enzyme called alcohol dehydrogenase. The enzyme converts the alcohol into a molecule called acetaldehyde, a toxin. Acetaldehyde, in fact, is ten to twenty times more toxic than the alcohol itself, and a carcinogen as well. When acetaldehyde builds up in the body, it destroys enzymes, protein, and DNA, and makes you vulnerable to many diseases, including alcoholic fatty liver disease and cirrhosis. Needless to say, your body has to get rid of it, and fast.

This is where glutathione steps in, helping to facilitate the second conversion of alcohol, transforming acetaldehyde into acetate, a water-soluble molecule you can easily eliminate from the body. In other words, you can now pee out the remnants of the alcohol, clearing your system. As I said, this is an efficient system for removing alcohol from the body, but it also takes a toll by using up a lot of your glutathione reserves. If you measure someone's GSH level after a bout of drinking, it will be close to zero.

Alcohol not only depletes glutathione in the liver, it also empties GSH stores in the lungs. The lungs are another area of the body where glutathione is typically found in fairly high concentrations; yet not so in chronic drinkers, whose lung glutathione levels become depleted, leaving them open to respiratory infections. People who abuse alcohol, in fact, are at an increased risk for pneumonia, and glutathione depletion is likely one of the causes.

There's one more thing I want to add about glutathione and its relationship with alcohol. Although alcohol may have some health benefits—there's some evidence that moderate drinking can help prevent heart disease—it can still tax the system in several ways. One of those ways is by acting as a solvent, dissolving toxins, such as organic phosphates, stored in the fat cells that might otherwise not be able to make it into the bloodstream. Alcohol opens the gate for these harmful substances, which will eventually end up in the liver—where they will need extra glutathione to get rid of them. This is something to think about before you drink as well as one more reason to do all you can to keep your GSH levels high.

CAN GLUTATHIONE CURE A HANGOVER?

THE ETERNAL SEARCH FOR a hangover cure has moved past the old "hair of the dog that bit you" to more sophisticated remedies. One of them is glutathione. Many IV nutrition spas and clinics use glutathione as a primary ingredient in their hangover treatments. The glutathione precursors alpha lipoic acid and N-acetyl cysteine (NAC) are also often recommended to relieve the aftereffects of a night of overindulging. Do any of these glutathione-boosters work? They're not miracle cures—so far nothing like that exists—but increasing the glutathione in your body can definitely help take the edge off a hangover.

If you've ever had a hangover, it needs no introduction. But hoping that you never have had the occasion to make a hangover's acquaintance, let's explore what it feels like and the biology behind it. Hangovers are generally experienced when you drink to the point of being drunk. Assuming intoxication leads eventually to sleep, you are likely to wake up with some of these unpleasant symptoms: fatigue, headache, increased sensitivity to light and sound, red eyes, muscle aches, nausea, and thirst. You might even have a rapid heartbeat, and feel tremulous, dizzy, and sweaty.

What's happening here? For one thing, dehydration. Alcohol suppresses a hormone that helps conserve fluids, leading the body to draw water out of the blood cells and dump it into the kidneys, where it's then released as urine. (It's no accident that establishments that serve alcohol must by law have restrooms.) And when you pee a lot, you not only lose water, you lose electrolytes as well. This may be why hangovers often come with headaches, fatigue, lightheadedness, and thirst. Alcohol can also disrupt your sleep patterns (another reason for fatigue), and the metabolites acetaldehyde and acetate, produced as your body burns off the alcohol, can cause headaches, sweating, nausea, and even vomiting.

Glutathione reduces the load of metabolites, particularly acetaldehyde, and in that way it does help alleviate hangover symptoms. But it can't do it alone. Dehydration is still one of the biggest culprits behind a hangover. IV therapies, although they may include glutathione, are largely hydration treatments, which is why they can be very effective. It helps, too, when an IV treatment includes B vitamins, since alcohol depletes them, contributing to hangover symptoms. If you're thinking about treating a hangover with alpha lipoic acid or NAC supplements rather than an IV treatment, great—just don't forget about rehydrating and replenishing your B vitamins too.

Because glutathione plays a critical role in alcohol metabolism, I think it's quite possible that shoring up your glutathione stores before indulging may help prevent a hangover. Having plenty of topical GSH on hand to quickly flush the acetaldehyde out of your body will undoubtedly lessen such symptoms as nausea and vomiting. You still might feel kind of lousy, just not quite as bad.

There's also another positive way that glutathione helps process alcohol: When you preload with glutathione before drinking, it can help you feel less tipsy. In fact, one of my patients, the owner of a large wine retailer, uses it to help him stay steady when he's hosting business wine tastings. Here's the story he told me:

"As the CEO of a major wine import company for forty-three years, you can imagine that I was drinking a lot of wine. My job was to travel the world and decide which wines to import to the US. (Yes, I had the best job in the world!) I would be tasting and assessing up to a hundred different types of wine, and sometimes Champagne, a day—which is a pretty interesting job for someone who traditionally does not drink alcohol at all and doesn't enjoy feeling buzzed or drunk. I had secretly always hoped that there was a magic bullet that could help me do my job well without getting intoxicated during these endless wine tastings.

"About seven years ago, I was introduced to a doctor that shared a topical glutathione solution with me. I had never heard of glutathione in my life, and I'm pretty sure I couldn't even pronounce it at first. But I was willing to try anything. I waited until my next wine tasting, and after about taste number fifty, I was definitely feeling the effects of the alcohol. Being concerned about doing my job professionally, I excused myself to the restroom and followed my doctor's advice to spray the topical glutathione on my abdomen. It was a bit sticky and smelt a bit like sulfur, but I didn't mind. I waited a few minutes for it to dry and then returned to the wine tasting area. It wasn't more than twenty minutes later that I felt about 95 percent back to normal. I couldn't believe it and wondered if I was having some sort of placebo reaction. But then I tried it again a few days later, and the same thing happened. I went to the bathroom, sprayed the glutathione, and bam!—within twenty minutes, I felt completely back to normal. It's like it was stopping the effects of the alcohol dead in its tracks. This was nothing less than revolutionary to me. The GSH solution became my new way of life, and a part of my work routine. And an added benefit was that after these extensive wine tastings, I would wake up the next morning with zero headache and zero hangover."

Even I was blown away by that story! Now, I wouldn't recommend that you use glutathione as an excuse to binge drink (and most certainly not an excuse to drink and drive), since its inebriation prevention can only go so far. But as the wine CEO has attested, it can help you feel sharper when you're drinking moderately. His story, of course, is unusual. Most people drink moderately and not nearly as often. Still, I think it also makes the case for taking all measures to keep your natural levels of glutathione high. If you are an occasional drinker, that's going to serve you well by making sure that your body can metabolize any alcohol that you do drink.

········•········

5 SLOWING DOWN THE AGING PROCESS

UMANKIND HAS BEEN LOOKING FOR THE FOUNTAIN OF YOUTH EVER since Ponce de León supposedly set out to find it in the 1500s. (The truth is, the search probably began well before his time.) It was a worthy endeavor then and it's a worthy endeavor now—nobody wants to grow old!—though actually remaining eternally young will probably elude us, well, eternally. But even if the fountain of youth hasn't materialized (no matter what late-night television ads may tell you), we *have* learned that it's absolutely possible to slow down the aging process. We've also come to know a lot more about the role glutathione can play in helping us maintain our health, vigor, and vibrancy as we grow older.

What is aging, anyway? Is it a state of mind? The state of the body? Or a combination of both? Aging is often just taken at face value and I mean literally at face value—how many lines and wrinkles you have, what sags here, what's sun-spotted there. To me, though, a big part of youthfulness is how you feel. When you're motivated to live life to its fullest, no matter what your chronological age, that's a sign of youthfulness. Of course, it helps if you are physically able to do the things you want to; that's where the state of the body makes a difference. Sometimes physical barriers to living the life you want can't be helped, but oftentimes having a youthful mentality can drive you to put in the effort to keep your body youthful too. I have seen ninety-year-old women do splits! Despite their advanced years, they are very, very strong because something within—determination—helped to make them stay that way.

By contrast, I have come across people who seem old because they've become increasingly uninterested in life. It may be because they find that their body can't do much anymore. Their joints hurt, they feel weak, they may even be hobbled by disease. Maybe they once had determination as strong as those ninety-year-old splits-performing acrobats I just described, but poor health has got them down. That's aging at its worst.

The truth is, our body is assaulted on a daily basis and the damage is cumulative. You cannot get rid of every single thing that contributes to the aging of the body—wear and tear on joints, sun exposure, stress, inflammation caused by free radicals—but you can give your body a fighting chance to heal itself and decelerate the aging process. That's where glutathione comes in.

Let's take a look at the various factors that drive aging, then explore the GSH connection.

WHAT CAUSES THE BODY TO AGE?

It won't surprise you to learn that we don't know everything about how and why the body ages. But scientists have offered many plausible theories, which generally fall into two camps. The first group of theories is based on the idea that we are programmed to age and eventually die. In other words, after birth, the clock starts ticking, driven by genetic forces that preprogram our obsolescence. The other overriding theories are related to the notion that simple wear and tear and the accumulation of damage cause us to age. Our body is like all machines (and cars, in particular) that eventually fall apart!

In all likelihood, how and why we age is multifactorial. *Many* changes to our physiology as well as nature's natural plan cause the deterioration we know as "aging." But I also believe that a major cause of aging can be pinned on something I've been talking about throughout this book: oxidative stress. We know that oxidative stress plays a role in the breakdown of the body. How big a role is still being debated, but many people — myself included — subscribe to what's called the free radical theory of aging.

In the 1950s, Denham Harman, a scientist at the University of California, Berkeley, developed the idea that aging was driven by free radicals' injurious effect on lipids, proteins, and DNA. Harman had first become interested in the topic of aging while reading a copy of his wife's *Ladies' Home Journal* magazine, and his interest followed him from his job working at Shell Oil, then through his years in medical school, and finally when he became a researcher at UC Berkeley. Over time, Harman believed, the wreckage left behind by free radicals led to the impairment of tissues and organs.

The way I look at aging is that, largely because of oxidative stress — the imbalance of free radicals and those antioxidants that quench them — our organs slowly become less capable of functioning at an optimal level as they used to do. They have lost the power to regenerate themselves in ways that allow them to work at peak performance. Take the lungs. When you are young, your lungs have the oxygen-carrying capacity of x. As you age that capacity diminishes, becoming $x - 1$, $x - 2$, $x - 5$, $x - 10$, and so on.

You can see this same sort of change most readily in your skin, the largest organ in the body. Young skin has several mechanisms in place — frequent cell turnover, the production of oil and collagen to maintain luster and structure, and more — to keep it looking fresh and plump. But the accumulation of damage, along with a biologically determined slowdown, causes the skin to function at a lesser level. This is aging you can see.

Same with age spots, the accumulation of melanin in your skin. Age spots are directly caused by oxidative damage, which in turn is caused by UV radiation — the sun. Again, this is aging you can see. But if there are age spots on the skin, are there also "age spots" on the liver, kidneys, lungs, heart, and just about everywhere else? I don't mean actual spots; I mean damage from oxidative stress. It's likely.

One way that oxidative stress does damage is by triggering inflammation in the body. Inflammation is a defense play by the immune system — it's part of the body's natural healing system, as you may have noticed anytime you've had a scrape or sprain. But not all inflammation is so readily apparent, and when inflammation inside the body becomes chronic, it can destroy healthy tissue, leading to many of the problems and diseases associated with growing older. You can't measure inflammation, but it's happening every single day of your life and your body needs to fight it with antioxidants. (See page 24 for more on inflammation.)

Over time, excessive and cumulative damage from free radicals and its by-product, inflammation, can contribute greatly to the compromised functioning and all-over slowdown we know as aging: aging you can feel, aging you can see, and aging you *can't* see (but will eventually make itself known to you through disease or disability).

THE TELOMERE CONCEPT OF AGING

Another theory, called the Hayflick Limit theory of aging, is based on the idea that cells can only divide forty to sixty times before they die off. The namesake of the theory was Leonard Hayflick, a scientist at the Wistar Institute in Philadelphia, who proposed the idea in the 1960s. One reason cells may die after a given amount of time is that their telomeres shorten. I mentioned telomeres briefly in Chapter 1, but let me repeat (and add to) what I said about them here.

Telomeres are most often described as caps composed of DNA and protein that sit on the ends of chromosomes (visually, they're often compared to the plastic caps on the ends of a shoelace). Telomeres protect the chromosomes in various ways and help prevent mutations in DNA. As the years go by and cells divide, telomeres become smaller. When they get small enough, they tell their cells to stop dividing, effectively hampering the body's ability to regenerate and refresh itself. This demise of cells is associated with aging as it weakens the structures of the body, which makes it more susceptible to aging, and can lead to organ failure. Shortened telomeres have even been associated with such diseases as cancer, heart disease, dementia, diabetes, and obesity.

Much of what we know about telomeres is thanks to the pioneering work of Elizabeth H. Blackburn, the Morris Herzstein Professor of Biology and Physiology in the department of biochemistry and biophysics at the University of California, San Francisco, and a Nobel Prize winner for discovering the molecular nature of telomeres.

One of the stated goals of Dr. Blackburn's lab at UCSF is to understand the mechanism of telomerase action and its various functions. The body has little protection against the shrinking of telomeres, but some cells have an enzyme called telomerase that can delay and even repair some shrinkage. Still, it can't hold off telomere-shortening completely. But would increasing telomerase in the body help reverse the aging process? That's a question Blackburn and others are looking at.

One thing we already know about telomerase is that glutathione is involved in its functioning. Researchers at the University of Valencia in Spain took a look at cell cultures in their lab and determined that glutathione regulated the activity of telomerase. So, when glutathione was high, telomerase was able to do its job at maximum capacity. The distinction here is that glutathione itself does not increase telomeres. But by enabling telomerase, it helps stop the degradation of the little chromosome caps, potentially extending the life of a cell. This is what's known as a secondary effect, and it's essentially clearing the way for other substances in the body to do their work.

Just as an aside, I'd like to tell you about some of the telomere-related work I'm doing in my lab. The latest buzz surrounding telomere lengthening involves a substance called cycloastragenol, an extract of the plant astragalus root. Given that cycloastragenol is the current front-runner in encouraging telomere growth, my research partner, Dr. Chinh Tran, and I went into the lab to see how this mysterious molecule played along with glutathione. We weren't sure how well they would synergize, but lo and behold, these two molecules were compatible. Maybe it's because they both perform reparative and/or regenerative cellular functions. I thought the melding of these two warrior molecules would make an interesting marriage: Could two telomere-lengthening substances work better than one? Time will tell.

The science surrounding telomeres and their role in aging is relatively young and we will undoubtedly learn more in the years to come. But already there are indications that adopting healthy lifestyle habits—many of the same lifestyle habits that also help you naturally increase your levels of glutathione—has an impact on telomere length. For instance, a small University of California, San Francisco, study found that men who, for five years, ate a plant-based diet (high in fruits, vegetables, and unrefined grains, and low in fat and refined carbohydrates); exercised moderately by walking thirty minutes a day, six days a week; reduced their stress levels through gentle yoga, breathing exercises, and meditation; and attended a weekly support group showed on average a 10 percent increase in telomere length. (And the more rigorously they adhered to the healthy lifestyle habits, the greater their telomere-length increase.) Men who were in a control group actually lost telomere length.

THE ANTIAGING ANTIOXIDANT

It's well-known that elderly people with chronic disease have low levels of glutathione. One group of researchers also wanted to know if the reverse was true: Do elderly people who are healthy and vibrant have high (or at least higher) levels of glutathione? They set out to find the answer, and the study that followed, published in the *Journal of Laboratory and Clinical Medicine* in 2002, was eye-opening.

The researchers who did the study were from the departments of biochemistry and molecular biology, pathology, and psychological and brain sciences at the University of Louisville in Kentucky. What made their work particularly interesting is that they not only looked at how glutathione was associated with physical aging, they also assessed if there was any relation to the subjects' satisfaction with life and overall sense of well-being.

The participants in the study were eighty-seven women aged 60 to 103, all in excellent physical and mental health as determined by a battery of tests. Over the course of five years, the women's blood was drawn three times to assess their levels of GSH. Their glutathione levels were also compared to those of healthy volunteers, who were between the ages of 20 and 39.

The ultimate findings of the study were quite remarkable: The healthy elderly women had levels of glutathione comparable to those of the younger healthy volunteers—and both groups had levels the researchers considered high. During the third wave of testing (five years from the start of the study), the elderly women's GSH levels dipped somewhat, but overall the researchers were able to conclude that healthy and mentally sound people have high glutathione levels despite their chronological age.

Does this explicitly mean that keeping your glutathione levels elevated will ensure that you age well and live long? It might. We are far from saying that GSH is the fountain of youth, but there are so many signs that it can indeed help us age more gracefully. As I discuss throughout this book, glutathione is a critical foe of oxidative stress. Given that oxidative stress is associated with accelerated aging, it follows that the antioxidant can help stem the tide. And there are other reasons to believe it may even help us live longer.

Some of this information comes from . . . bugs. Certainly, anyone who's ever swatted a fly without giving it a second thought would be hard-pressed to see that we have much in common with insects, but the truth is, we do share many traits. Mosquitos, in particular, have biochemistry that's similar to humans', including the fact that their glutathione levels drop as they age. This fact set the stage for a study by some of the same Louisville researchers who looked at the blood GSH of healthy elderly women. In this study, they fed yellow fever mosquitoes a solution that increased their glutathione levels by 50 to 100 percent. These mosquitoes typically have a short life span of thirty days. With the added glutathione in their diet, the median lifespan increased to forty days, a 38 percent rise.

Another insect study offers further insight into glutathione's potential for life lengthening. This study, done in Switzerland, involved *Drosophila melanogaster*, more commonly known as the fruit fly. The researchers approached the investigation with the idea that glutathione may not only slow aging by deterring oxidative stress, but also by positively affecting gene activity. To increase the fruit flies' GSH levels, they were fed N-acetyl cysteine (NAC), which, by providing the building block cysteine, helps the body produce more glutathione. The results proved the researchers' hypothesis right. Fruit flies fed 1 mg/ml of NAC food lived 16.6 percent longer; those who got 10 mg/ml had a life span increase of 26.6 percent. What's more, the glutathione prevented age-related changes in RNA (the messenger acid that carries DNA messages for protein building) that, like oxidation, may influence life span.

Here's one way to think about glutathione and aging. You know what they say about planes? A plane can fly hundreds of thousands of miles for years without any difficulties, largely because all of its components are in use and it receives regular service. But if you put that same plane in a hangar for a year, then try to fly it, you're going to have some problems. Bodies, like planes, can get "rusty" when there's not active upkeep. Glutathione is constantly removing the rust from our body, keeping it in good working order for years and years to come.

CAN GLUTATHIONE HELP PETS LIVE LONGER?

WE ARE STILL LEARNING about what glutathione can do for pets, but the forecast is good. What I find intriguing is that published research shows that, just as in humans, dogs with illnesses are low in glutathione. Using GSH, it seems, would be a good way to help pets fight disease and other conditions that shorten their life.

I'm still gathering stories about what happens when people treat pets with glutathione, but so far what I do know is that pet owners (dog owners, specifically) who have tried topical glutathione are convinced that it's improved the overall health of their pets. Since better health often translates into greater longevity, GSH may even help pets live longer. More research is needed, but in the meantime, if you want to try glutathione on your pet, you can feel comfortable that it's safe. (My own dog Auggie loves to lick glutathione off my hand—it's the sulfur smell so unappealing to us and so appealing to dogs!) Also note that pets can benefit from ways that help them build glutathione themselves, such as by adding whey protein (see page 140) and N-acetyl cysteine supplements to their diet.

6 ENERGY, EXERCISE, AND GLUTATHIONE

I F I HAD TO SUM UP ALL THE INFORMATION IN THE PREVIOUS CHAPTERS, I would say this: Glutathione simply makes everything in the body operate better. And if everything is operating better, you're going to feel better. That's why, I believe, so many people who improve their glutathione status find that their energy level improves right along with it. Glutathione is also integral to the processing of oxygen in the mitochondria, ground zero for energy production in the body. When mitochondrial glutathione is high, it's likely that you're going to feel vibrant and less prone to fatigue.

One area related to energy that I want to explore now is exercise. There's plenty of evidence to suggest that low- and moderately paced exercise can reduce fatigue and increase vigor, including a University of Georgia study that found students who complained of being tired all the time felt more energetic after working out for thirty minutes, three days a week, over a span of six weeks. And, of course, there's no doubt that physical activity improves health overall. But maybe you've heard this too: exercise increases oxidative stress, an indication that it depletes glutathione in the body. This brings up an important question about exercise as it relates to GSH: If you're trying to increase glutathione in your body, should you still get in your daily run (or swim, or bike, or yoga class—whatever it is that you do to stay fit)? The answer is yes, and I'll explain why.

EXERCISE AND OXIDATIVE STRESS

It's well-known that exercise, particularly prolonged or short but intense aerobic exercise, increases the production of free radicals. When you work out, your metabolism increases and you burn more calories and fat; free radicals get created in the process. Exercise also increases the amount of oxygen you take in, another reason free radical production rises. If you're, say, running in the park or pushing hard on an elliptical trainer—anything that makes you breathe faster and deeper—your oxygen intake might increase anywhere from 5 to 6 liters a minute to 100 liters a minute. As your body converts that increased oxygen into energy, it releases free radicals looking to steal electrons from healthy cells and tissues. And the more vigorous your workout, the greater the number of free radicals you generate.

Is this bad? Yes and no. Oxidative stress, of course, can make the body more susceptible to illness and disease by killing off or altering healthy cells. Oxidative stress can also interfere with the quality of exercise and its aftermath by increasing muscle fatigue and postexercise muscle pain.

And yet, some research suggests that the free radicals generated during exercise also have a positive role to play in the body. In fact, they may figure into the process by which we adapt to exercise. With training, our body "grows" mitochondria (more mitochondria, more energy) and our muscles get stronger. Free radicals, it seems, can help trigger these changes. What's more, although intense exercise may generate unhealthy levels of oxidative stress in beginners, as you become better trained, your cells get better at handling larger waves of free radicals. And they do that by producing greater numbers of endogenous (made-by-the-body) antioxidants, most especially glutathione. Exercise, in other words, helps boost the manufacture of glutathione and also creates its more efficient release from the tissues.

There is some evidence, too, that if you exercise throughout your life, you will maintain higher glutathione levels as you age. In 2016, at the annual American College of Sports Medicine meeting, Spanish researchers presented a paper on a study involving twenty-three senior men, divided into groups according to their lifelong level of physical activity. One group was sedentary, one group was moderately active, and one group comprised amateur endurance athletes. Of the three, the amateur athletes had the highest levels of glutathione and the lowest indicators of oxidative stress.

As I talked about in other parts of this book, balance in the human body is key, and so it is with exercise and its effect on the free radical–antioxidant equilibrium. The truth of the matter is, sports science experts are still trying to figure out the harm-to-benefit ratio when it comes to free radicals and antioxidants, but we do know for certain that exercise, and plenty of it, helps prevent disease, so even if it creates some oxidative damage the overall effect is positive. We know, too, that regular exercise helps improve the free radical–antioxidant balance, all the more reason to weave workouts into your weekly schedule.

MUSCLING INTO ATHLETICS

ONE REASON *GLUTATHIONE* HAS become a buzzword in the athletic community is that it decreases one of the biggest enemies of performance: fatigue. Stephanie Hunter has seen this firsthand. Hunter, who founded the Hunter Method, a muscle therapy clinic in Jupiter, Florida, frequently works with athletes and other people in high-powered jobs, such as superhuman Tony Robbins, who test their physical limits every day. She uses glutathione as part of their treatment for pain and/or recovery. "The most important factor in performing at high levels is not how high you can jump, but how quickly you can get your muscles to heal," says Hunter. "GSH speeds up recovery time and decreases both general and muscle fatigue. The body is an amazing healing machine, and GSH seems to be the main motor of that machine."

All exercisers can benefit by ensuring that their glutathione "cupboard" is fully stocked, but it's particularly important for high-intensity athletes. My colleague Jim LaValle, RPh, ND, a clinical pharmacist and certified nutritionist, who is also program director of integrative medicine for the NFL Hall of

Fame Health and Performance Program, notes that as GSH goes down, the ability to maintain a healthy cellular environment goes down right along with it. "And that's important for the athlete that's performing." Glutathione, he adds, helps clean up acids, such as lactic acid, produced during vigorous physical activity. "If you have an adequate glutathione network in your body, you're going to be able to clear the tissue damage that occurs due to intense exercise."

........•........

THE GLUTATHIONE EFFECT

There is a lot of research to suggest that glutathione is on the forefront of the body's defense against exercise-induced oxidative stress. When, in studies, subjects are put through the paces—those paces ranging from cycling to doing strength-training arm exercises—and then their oxidized glutathione levels are measured, it's clear that significant GSH has been "used up" to battle free radicals.

So, what would happen if you gave extra glutathione to exercisers? Would it improve their response to physical activity? If there's one thing that any exerciser wants—whether that exerciser is a three-day-a-week gym-goer or a professional athlete—it's to feel less fatigue as they work out and to quickly recover from the exercise. Glutathione has been shown to help on both accounts in many different studies. Some research, for instance, has found that when exercisers (both human and animal) are provided with supplements that increase glutathione (generally the cysteine-carrying supplement N-acetyl cysteine [NAC]), it improves their endurance.

Let's look closer at one of those studies, one that had well-trained endurance athletes cycle semivigorously on a stationary bike for forty-five minutes, ending with a push that took them up to the highest intensity they could achieve (a sprint to the finish, you might say). In this study, conducted at Victoria University of Technology in Melbourne, Australia, the eight young (median age was twenty-seven) men who participated were very fit. Before the research took place, they had been either running or cycling four to five times a week for one to two hours and had been at it for at least two years. During the study, they came into the lab and cycled on six occasions, each exercise bout separated by seven days. The men received NAC intravenously both during and after exercise.

By the study's end, it was clear that the NAC boosted their glutathione levels and that, possibly by using up that extra glutathione, the cyclists improved the time before they fatigued by an average of 26 percent. I say "possibly" because it's difficult to say for sure what improved the exercisers' endurance, but since their blood tests showed more oxidized glutathione in their blood, I believe it's possible to deduce that the GSH helped prevent the oxidative damage that can lead to fatigue. Another study, this one done on animals, seems to drive home this point as well. When exercising rats were treated with a GSH-depleting serum, their endurance dropped by 50 percent.

In the real world, I doubt anyone is going to have an infusion in the middle of a workout, and yet these studies and others like them do tell us something important: Raising glutathione levels also raises endurance, most probably by reducing oxidative stress. I believe there may also be a couple of other ways that glutathione improves endurance in exercisers. One of them is by enhancing lung capacity. Because there is considerable glutathione in the lungs, GSH can literally help you breathe easier (inhaled glutathione has even been used to treat certain lung conditions). Another factor in endurance is muscle repair. Exercise involves the process of breaking down and rebuilding muscle—that's how we get stronger. Glutathione assists in this biological activity as well.

In my own work, I've had the opportunity to see how glutathione can benefit people who work out hard. Several athletes have used our topical GSH to help them perform at the top of their game, recover more quickly from their workouts, and stay fired up. Since glutathione is not a performance-altering drug but classified as a topical vitamin, it is a welcome nutritional supplement in athletics. In fact, my company has begun conversations with the medical divisions of the NFL and many other professional and amateur sports leagues to supply topical glutathione to be used in their training and recovery protocols. I have also been invited to supply topical glutathione to select Olympians who are in training.

The excitement about glutathione in the athletic community is understandable. Increased glutathione availability, whether it comes from supplementation or through the natural means you'll learn about in this book, restores the body's inherent ability to repair itself and create energy. It removes the sluggishness, like taking a wet blanket off your body. Glutathione, in short, allows your body to do everything it wants to do, whether you're a linebacker or a mom trying to keep her kids in line!

A HORSE STORY

TO TELL THE TRUTH, the racehorse's owners didn't expect too much from her. Built like a brick house, Flame N Flash was a steady ride, but not really the type of horse to make headlines in the *Daily Racing Form*. And yet she qualified for the Mildred N. Vessels Memorial Handicap race at Los Alamitos Race Course in California, the leading quarter horse track in the country. "We didn't have high hopes for her," says Tamara Sofi, a nutrition expert and holistic practitioner who was part of Flame N Flash's training team. "Mares are like mean girls and there were some intimidating horses in that race."

What happened next took everyone by surprise. Flame N Flash led the race the whole way. And the $150,000 purse she won gave her a berth in the Champion of Champions race five weeks down the line. But first there was the Super Derby trials in two weeks, where the filly would be up against the boys. "Colts usually win that race," says Tamara. Not that year. Flame N Flash won again. When she ultimately ran the Champion of Champions race, Flame N Flash, the youngest horse and the only filly, led the whole way but was beat out by a nose. Her performance, though, was enough to get her crowned three-year-old filly of the year, and up her desirability for breeding. "People were calling up and asking, 'What's going on with this horse?'" says Tamara.

If you think glutathione might have something to do with this horse story, you're right. I knew Tamara, an expert in free radical science, through a professional connection. When I heard she was working with racehorses, a lightbulb went on. Could glutathione improve a horse's performance? I already knew many athletes who used glutathione to improve their stamina. I asked Tamara whether she wanted to give it a try, and she had the perfect candidate: Flame N Flash.

The racing barn where Tamara works forbids the use of animal products and, of course, can't use drugs or supplements banned by the racing association. No problem. Our GSH solution is vegan, track-cleared, and the delivery system was simple. Tamara simply sprayed it on Flame N Flash's belly, beginning two weeks before the Vessels Handicap race. It's necessary to discontinue use of a lot of substances long before race day, but the glutathione is so natural that Tamara was able to apply it as close as two hours before a race.

"The way I think the glutathione made the biggest difference was in recovery," she says. "There were only two weeks between the first two races, then the Champion of Champions was three weeks later. That's a quick turnaround. It's easy for horses to win in the beginning of the year, but to have them continue winning is more difficult. Antioxidants are critical for recovery."

I agree with Tamara that improved recovery was the main benefit of the glutathione. I think glutathione's ability to improve lung capacity and enhance muscle repair probably helped Flame N Flash as well.

Since the horse's success, we've had the opportunity to continue exploring the benefits of glutathione for racehorses. We've been approached by horse owners at the Del Mar track in San Diego who want to try out glutathione on horses who've had health challenges. I'm optimistic that we will learn a lot more about glutathione from working with these big, beautiful animals.

········•········

7 REJUVENATING SKIN

A FEW YEARS AGO, I WAS INTRODUCED TO A GROUP OF WOMEN WHO were avid tennis players. They had been playing the game out in the sun for years and years, and you could see the consequences of all that exposure in one particular area: their décolletage, laid bare to UV rays by the women's preference for V-neck shirts. Compared to other areas of the body, the skin on their chest was red and wrinkled—even with the use of sunblock. Another tennis player, the longtime champion Jimmy Connors, presented a similar problem to me, only it was mainly his arms that had seen the brunt of all those hours out there trading ground strokes with John McEnroe.

The advanced aging of these outdoor athletes' skin is a prime example of what's called extrinsic aging. Extrinsic, meaning "coming from the outside," refers to aging precipitated by the elements, primarily the sun but also pollution and smoke. This type of aging—known as photoaging—causes wrinkling, discoloration, pigmentation spots, and coarseness, among other visible effects. The skin is also subject to intrinsic aging, the natural process of tissue degeneration, which piles on with a breakdown of collagen and elastin (elements that give skin plumpness and "bounce-back"), skin thinning, fine lines, and slower cell turnover. Typically, young skin loses about forty thousand cells a day, keeping it looking radiant. But then, things slow down. The skin refreshes itself every 14 to 21 days in your late teens and twenties. Fast-forward to your thirties and now it will take more like 28 days. In your forties and fifties, it slows further to a 45- to 60-day turnover rate, then further still—to 60 to 90 days—as you reach your sixties. With all those dying or dead cells hanging around, older skin just doesn't look as vibrant.

Genetics have a lot to do with the intrinsic aging of skin, but it's helped along by oxidative stress. I know I must be beginning to sound like a broken record, but oxidative stress really is at the heart of so many physiological issues. Free radical production increases as we age and our body's ability to repair DNA fractured by oxidative stress declines, leaving the skin vulnerable if it is not protected by ample amounts of glutathione and other antioxidants. Oxidative stress is also a factor in extrinsic aging. The skin's exposure to toxins in the environment boosts free radicals, and so, of course, does exposure to UV rays. In fact, the type of free radicals generated by ultraviolet radiation is particularly toxic.

To emphasize just how much the sun can affect the skin, in 2012 the *New England Journal of Medicine* published a remarkable photograph of a sixty-nine-year-old man. The man had been driving a truck for twenty-eight years, exposing one side of his face (the side closest to the driver's window) to the sun's rays far more than the other side. After all those years of driving, the differences between the two sides were stunning. It was almost as if the truck driver was two different people. On the sun-exposed side, his skin had a coarse texture, a thick web of wrinkles, and sagged. The side turned away from the window was only slightly lined. It was comparing a plum to a prune. While the effects weren't quite as dramatic, the disparity in the overexposed and unexposed skin of the tennis ladies and Jimmy Connors was similar evidence of just how much damage the sun can do.

FORM FOLLOWS FUNCTION: WHAT MAKES SKIN LOOK AGED

THE HALLMARKS OF AGING—LINES, wrinkles, less uniform color, sagging, dryness—have biological underpinnings. Here are some of the changes that affect appearance most.

<u>Thinning</u>—The thickness of the skin declines 6.4 percent per decade on average.

<u>Lowered cell turnover rate</u>—In your teens and twenties, skin may turn over as rapidly as every 14 days; in your senior years turnover can be as sluggish as only every 90 days.

<u>Less active melanocytes (pigment cells)</u>—A drop of 8 to 20 percent per decade makes skin color become uneven with age.

<u>Fat loss</u>—Older people have as much as 65 percent less fat in the skin, causing sagging.

<u>Reduced sebum production</u>—The oily substance that protects and keeps the skin luminous can drop by 60 percent in older women.

<u>Decreased collagen and elastin</u>—Collagen decreases about 1 percent with each passing year; by age 40, elastin production slows to a crawl. This affects the "architecture" and bounce of the skin.

........●........

GLUTATHIONE'S COMPLEXION CONNECTION

Glutathione plays a critical protective role in the skin. When your body is hit by UV rays, glutathione is there at the ready to keep damage to a minimum. Researchers have determined this fact a few different ways. There have been studies, for instance, that explored what happens when skin cells in a lab setting are depleted of glutathione, then exposed to UVA and UVB rays. (UVA rays are longer and are associated with skin aging. Shorter UVB rays are associated with sunburn. Both do the kind of damage that can lead to skin cancer.) In the absence of GSH, the cells become much more sensitized to the light, an indication that the antioxidant is an important shield against sun damage. One study also looked at glutathione levels in hairless mice exposed to UVA and UVB light, only to find that the rays depleted GSH. The dwindling amount of glutathione in the cells shows that, when the sun shines in, glutathione goes to work.

It's not known whether keeping your glutathione levels high through natural means can help guard against sun-induced aging of the skin or skin cancer. That said, common sense dictates that it's important to have all hands on deck—the more glutathione you have in your system, the better. And we do know that supplementary glutathione, especially used in conjunction with other complexion-enhancing antioxidants, can protect and visibly improve your skin. Can it totally turn back the clock? Nothing can. But supplementary glutathione can help stop the damage and even restore some of the skin's youthful vitality.

SKIN BRIGHTENING AND SMOOTHING

One of the ways supplemental glutathione improves the skin is by lightening redness and age spots, both caused by sun exposure. To understand how it works, it helps to know a little about how skin reddens and darkens beyond its usual color in the first place. In other words, how and why do you get a tan or a burn? Or, in the case of an already brown or black complexion, what makes skin color deepen? What causes freckles and age spots?

When rays penetrate the skin, they reach down below the surface where specialized cells called melanocytes live. Melanocytes create a pigment called melanin, which darkens the skin and has the ability to absorb UV light. So, melanin isn't just there to give white skin a glowing tan; it's actually the body's way of trying to protect the skin from further insult. Without this protection in place, the skin will burn, a reaction that occurs when rays damage skin cell DNA, prompting chemical reactions that lead to inflammation. Unfortunately, by the time skin darkening occurs, it can be a little like closing the barn door after the horse has escaped. Some harm may have already been done to the skin cells; extensive damage can even lead to cancer. There can also be several adverse cosmetic effects. Over time and with frequent sun exposure, melanin can clump together, creating freckles and age spots.

The way UV light kicks off the production of melanin is by sending an enzyme called tyrosinase into action. Glutathione suppresses the activity of tyrosinase, preventing the skin from darkening. The actions of glutathione also help to convert melanin to a lighter color so that age spots become less apparent.

More and more people have begun using glutathione to address hyperpigmentation. Some treatments involve IV glutathione; some are oral; others are topical. Whether you apply a glutathione solution directly on the affected spot or just get it into your system (by, say, applying it to your abdomen, a good place for absorption, or through an IV), it's possible to see a real difference in skin that's been treated with glutathione (see "Skincare's New Star," page 112). And some research has also born it out.

In one study, conducted by medical researchers in Thailand, sixty women were given one of two different forms of oral glutathione supplements or a placebo. The analysis of skin on their arms and faces found that, after twelve weeks, melanin was lower in both of the glutathione-treated groups. They also had reduced wrinkling and better skin elasticity. (And this was using oral supplements, which aren't even absorbed that well—I'd imagine the results would have been even better with a more bioavailable form of glutathione.)

An earlier study by a different team, this one in Japan, had similar results. This time, the researchers recruited thirty women and had them apply a glutathione lotion to one side of their face and a placebo lotion to the other side twice daily for ten weeks. They then measured the skin's melanin, moisture content, smoothness, wrinkle formation, and elasticity. Melanin was significantly lower on the glutathione lotion side of the face. In fact, everything was better. The skin was moister, smoother, and had fewer wrinkles.

Although I can't say for sure what accounts for the better condition of the skin after glutathione supplementation, it's most certainly related to GSH's power to reduce oxidative stress. Take away those free radicals and you're less likely to have visible signs of aging.

One thing that's interesting about using glutathione to rejuvenate the skin is, as I mentioned earlier, it can work systemically—that is, while applying it to the trouble spot (as the researchers did in the Japanese study) can be very effective, getting more GSH into your system can also give your skin a boost. I heard from a woman named Janney, who began using glutathione solution upon the recommendation of her naturopath doctor after being diagnosed with adrenal fatigue and leaky gut syndrome (a condition that involves bacteria and other toxins permeating the intestine walls). It had made her extremely lethargic, mentally foggy, and caused her all sorts of digestive issues.

That was what was going on inside. On the outside, Janney's skin had turned quite discolored and was dotted with age spots. She began a regimen of applying GSH twice a day on her abdomen and once a day on her face. She was hoping the treatment would transform her overall health, and it did, helping to resolve her digestive issue and increase her energy. What she wasn't expecting was that her skin would become much brighter, smoother, and less wrinkled. I think part of Janney's skincare success was that she was using glutathione not only where it was needed most (her face), but also where it could be absorbed into her system, making more GSH available to fight oxidative stress in the skin.

SKINCARE'S NEW STAR

It had been a few months since I'd last sat down with tennis legend Jimmy Connors. I received a last-minute phone call to meet him down at a tennis club in Carlsbad, where Jimmy and Spencer Segura (son of the tennis great Pancho Segura) were working at a charity tennis event for underprivileged children. He wanted to discuss how glutathione could treat the cellular damage that he'd suffered due to continuous exposure to the sun.

We were sitting outside on a particularly hot and humid day, 97 degrees Fahrenheit, to be exact, and I noticed that despite the brutal heat, he was wearing long sleeves. "Why?" I asked, and he showed me the skin on his arms. Of Irish descent, Jimmy was particularly vulnerable to UV rays, and the years outdoors on the court had taken a noticeable toll, most visibly on his face and forearms.

A few weeks later, I got a message from Jimmy that the glutathione solution and cream I'd given him were yielding results. "Wow, this stuff really works!" he said. We had similarly great results with the tennis ladies who saw the redness and age spots on their chest (décolletage area) start to go away after only a few weeks.

Glutathione is increasingly becoming a marquee ingredient in many skincare products, prized for its ability to heal, beautify, and protect. It's long been known that antioxidants, such as vitamin C, play a key role in helping build collagen and repair sun damage. We knew about glutathione, too, but getting it down into the deeper reaches of the dermis where it could get to work has not been easy. When Dr. Tran and I came up with a form of GSH that is easily absorbed into the skin, it seemed natural to add it to skincare products, and that's how our Auro Skin Protocol line was born. (For more about Dr. Tran, see "Meet My Partner," page 114).

Creating a skincare line is a lengthy process. We knew that, as much as glutathione is the mother of all antioxidants and a master at subduing free radicals, it could use help from other compounds as well. Wrinkling is a multifactorial problem, a result mostly of a breakdown of collagen and elastin. So, we set out to make a complex that included other active ingredients, including vitamin C, coenzyme Q10, carnosine, and resveratrol.

Resveratrol is a group of compounds called polyphenols that have antioxidant powers. It's most famously an ingredient in red wine. The average recommended dose for resveratrol is 1,000 to 2,000 mg a day. It takes about forty-one glasses of red wine to equal 20 mg of resveratrol, so in doing the math, you would need approximately 3,075 glasses of red wine a day to get your suggested daily dosage. That's laughable. So, we decided to put a high concentration of it in our GSH antiaging creams instead.

Originally, we also included alpha lipoic acid in the formulation, but found that this potent antioxidant did such a good job of detoxifying the skin that it caused rashes, itching, and redness. The skin was trying to get rid of stuff! It took a while to figure out which ingredient was causing the problem; when it turned out that alpha lipoic acid was the culprit, we took it out.

This gives you a little insight into how skincare products are born. Fortunately, glutathione is now getting its due as an important ingredient in fighting aging and brightening and repairing the skin. You'll likely see more GSH-based lotions, creams, and serums come onto the market in the future (I am, of course, partial to our lab's!) and finding the right one may just be a matter of trial and error. Changes won't happen overnight, but if the glutathione in your product is working, you should see changes in about a month.

One person who's had great success with glutathione is Ali Landry, an actress, model, and former Miss USA. "I love skincare, I always have. While growing up, I sold skincare products out of my mother's beauty salon to put myself through college," Ali told me. "After winning Miss USA and then becoming an actress, I was introduced to some of the finest estheticians and skincare professionals around the world. I've tried just about every skincare brand in the market, but what I look for are 'wow' products, true game changers, that are results driven."

When Ali tried using our glutathione product, the first thing she noticed was that her skin appeared thicker. "That made it look more youthful," she said. The GSH also softened fine lines and wrinkles on her face after about two weeks. After using the products for eleven months, her sun spots and melasma (and even scarring from a blemish) diminished by about 85 percent. "It's now in my collection of 'must-haves,'" said Ali.

MEET MY PARTNER

THE PERSON WHO HAS pioneered most of the cutting-edge formulas that come out of our lab is Dr. Chinh Tran, the co-inventor of our transdermal glutathione and subnano delivery mechanism. I generally refer to him as my research partner, but he is infinitely more than that. Not only is he one of the most deeply methodical chemists that I've ever known, and a fellow alum from the University of Southern California, but he has become like a brother to me. Chinh is the epitome of the unsung hero and mastermind behind the scenes.

A political refugee from Vietnam, Chinh escaped a volatile situation and fled to the United States at age twenty-six. To my delight, he ended up at USC, where he reestablished himself by studying the American curriculum. He went on to be one of USC's prestigious research and development (R&D) partners for twenty-five years, which speaks volumes about Chinh's capabilities. Once I had the chance, I was wise enough to snatch him away from the university and bring him over to my compounding pharmacy business as my R&D leader.

Chinh received his PhD in physiology and biophysics at USC in 1993. His research has been published in many international journals, such as the *Journal of Biological Chemistry,* *Biochemistry,* and *Biophysical Journal,* on such topics as the structure and function of various important biological enzymes. I admire his professional accomplishments, but his integrity and humility are what I most respect about him. It's commonplace for Chinh to bring me something groundbreaking he's discovered in the lab, and it's this wonderful and endless Disneyland of scientific adventure that drives us both. I feel fortunate to have found the perfect business partner, as well as someone committed to creating products that help the human body do what it is naturally supposed to do.

························•························

SUNBURN, BRUISING, AND SCARS

If you've ever had a severe sunburn you may have not only felt sensitive to the touch, but also felt headachy, nauseous, and chilled. This is because a burn can be more than skin-deep, both doing local tissue damage and affecting what's going on throughout your system. Sunburns increase free radical activity, overwhelming the antioxidants at work in your body.

On a smaller scale, the same thing happens when you obtain a bruise. The injured part of the skin releases free radicals, and can even cause melanin formation, increasing pigment in the affected area. This same increased melanin response is also the reason some injuries and cuts turn into scars and stay a darker color than the rest of your skin.

Increasingly, glutathione is being used to alleviate the damage of burns and skin injuries that lead to melanin formation and scars. With burns, topical GSH reduces the pain almost instantly and helps speed the recovery process. Even when the damage is part of a cosmetic procedure, glutathione can help. Laser procedures, for instance, are purposely designed to traumatize the skin so as to promote a healing response that includes a boost in collagen production. GSH is effective in helping laser burns to heal quicker because it rushes in to quench all the free radicals being released by the skin. More cosmetic surgeons are also turning to topical glutathione as a trusted resource for reducing scarring from surgical procedures.

THE GSH KEY TO SKIN DISEASES

If you have psoriasis or know anyone that does, you are aware of how uncomfortable and traumatizing it can be. Extra cells build up on the surface of the skin, creating red, itchy, and scaly patches. There are even more severe forms of the disease that cause blisters and even arthritis. The exact cause of psoriasis has not been pinpointed, but it seems to be related to a glitch in the immune system. (For more on autoimmune diseases, see Chapter 3.) It's often inherited and can be triggered by all kinds of things, including stress and infections. There really isn't a cure for the disease, but it can be managed so that flare-ups are rare.

And what is the glutathione connection? Psoriasis is among several skin disorders—including eczema, dermatitis, and lupus—that are linked to low levels of glutathione. One study found that the more severe the symptoms of psoriasis, the lower the GSH activity in certain white blood cells. Without glutathione available to detoxify the cells, enzymes that lead to psoriasis symptoms get free rein.

In 2013, a group of researchers from George Washington University in Washington, DC, and McGill University in Montreal took a look at how well psoriasis sufferers fared when treated with glutathione. Specifically, they had seven patients take 20 g of nondenatured bioactive whey protein isolate per day for twelve weeks. (I'll tell you more about whey protein in Action Plan One, but briefly, it's a derivative of milk that contains all the amino acids needed for glutathione production.) The patients' psoriasis varied in severity and they were already treating it in different ways. They were told to continue their treatments (which ranged from topical creams to light therapy to steroids). By the end of the study, and no matter what therapy they were already using, all the patients had visible improvements in their skin. While the other treatments may have been doing some good to relieve the patients' psoriasis, glutathione seemed to take skin repair a step further. It was a very small study, but the results were certainly impressive.

ATTACKING ACNE

RECENTLY, ONE OF THE most avid fans of glutathione I know told me a funny story. Michele was sound asleep, getting in a few extra minutes of sacrosanct weekend shut-eye, when she was awoken by the unmistakable sound of giggling. Lifting up a corner of her sleep mask she could see her teenage daughter, Brighton, and two friends down on the floor, stealthily army-crawling their way toward the bathroom. "Shhhh . . . Hey you guys, be superquiet! My mom told me she only has a little glutathione left. She'll freak out if she knows we're using it."

Michele stifled a laugh. She'd been using GSH to help soften the lines and lighten the age spots that were beginning to appear on her face. When Brighton tried it on her acne, she, too, had great success and, voilà, a glutathione burglar was born. Michele could hardly blame her daughter for wanting to share the magic with her friends, so she kept quiet as the girls grabbed the GSH, army-crawled back out of the room, and ran down the hall as if they'd just stolen gold in a bottle. And, in a way, they had.

Acne—the catchall term for blackheads, whiteheads, and pimples—occurs when the pores get plugged with a combination of oil (sebum) and dead skin cells. Sometimes, as in the case of pimples, they become inflamed with bacteria. The inflammation along with all the clogging (and sometimes pus) is what causes bumps. When the skin becomes infected—which essentially is what happens when there's bacteria involved—the immune system slides into gear. In the process, it generates free radicals, exacerbating the situation by causing more inflammation and redness.

In the case of papulopustular acne, a severe form of the condition, research shows that the antioxidant defense system, including a form of glutathione, may be impaired. Some research also indicates that a decrease in GSH may contribute to regular forms of acne. In 2011, Japanese researchers collected skin samples from people both with and without acne, then compared the quantity of glutathione in each group. The volunteers were all women between the ages of eighteen and thirty-two, and the researchers looked at glutathione in samples from both their face and their upper arms. The amount of GSH in both areas was significantly lower in the women with acne than in the women with clear skin. Interestingly, there was low GSH in the skin on the upper arms of the women with acne, even though it wasn't a part of the body that suffered from acne. This suggests that simply having low levels of glutathione overall—not just in the affected area—can predispose the skin to acne.

So far, we haven't seen studies that look at how increasing your glutathione levels through healthy lifestyle habits can affect the skin, but it doesn't take much of a leap to think that it may help with acne. I can say with certainty that *topical* glutathione can help heal acne, having seen it in my own patients. The combination of the two, well, I believe that it's the ticket to clearer skin. When you have glutathione coming at the problem from every which way, you are more likely to get a good result.

········●········

PART III

Putting Glutathione to
Work—3 Action Plans

8 HOW DO I GET MORE GLUTATHIONE?

THE PURPOSE OF THE THREE ACTION PLANS THAT FOLLOW IS TO HELP you increase your levels of glutathione. Each one lets you zero in on a particular goal, whether it be just to improve your overall health (Action Plan One), detox (Action Plan Two), or rejuvenate your skin (Action Plan Three), with all of the plans building on the benefits of Plan One. The overriding objective of all of them is to help you optimize the natural production of glutathione in your body. However, at times, you may find that achieving the results you desire necessitates turning to supplementary GSH, especially as you age and glutathione production slows down.

It's no secret that I'm in the glutathione business, but that doesn't mean that I want to leave you in the dark about all the different kinds of products out there except mine. To my mind, the more you know about glutathione supplementation and the more options you have, the better. So, before you dive into the Action Plans, take a few minutes to read up on how that tricky antioxidant/detoxification molecule gets delivered from the lab to your body. I'll tell you about how and why I developed a glutathione solution and fill you in on other forms of supplemental GSH.

THE VARYING FORMS OF GLUTATHIONE

Glutathione has been the most researched supplement online for the last three years in a row, and there are over 800,000 searches on Amazon each month for different glutathione products. Clearly, many, many people want to try the antioxidant. But all delivery systems are not equal. The primary reason is that GSH is a very bulky, unstable, and reactive molecule. It's unruly and aggressive by nature, because it's supposed to be: Once inside the body its job is to seek out all of the bad guys (free radicals), so of course it's going to be highly reactive. But these characteristics have also made creating a stable form of GSH that is well absorbed by the body one of the great challenges of the pharmaceutical world. Perhaps the reason most people haven't heard of glutathione yet is because there hasn't been a good therapeutic, easy-to-use delivery system until recently.

This was the challenge my research partner, Dr. Chinh Tran, and I undertook when we began working with glutathione. We discovered in the laboratory that we were able to manipulate the actual glutathione molecule in order to keep it in a reduced (meaning that it's ready to stabilize free radicals) and water-soluble form. This way, GSH can be delivered through the subnano water channels (pores) of the skin via a simple topical solution. The water channels of the skin are tiny, but act as a highway, allowing for rapid absorption. The glutathione in our transdermal products is produced by a pharmaceutical company via a fermentation process. Since 2009, GSH has been classified by the FDA as GRAS (Generally Regarded as Safe). Our topical glutathione is identical to how our own body produces GSH, using the same three amino acids, in the correct sequence.

All of our initial studies have been incredibly promising, and the results and testimonies from everyone using the topical glutathione solution have been nothing but positive. Some of the results have been shared in this book and are quite groundbreaking. The subnano delivery mechanism and the science behind it have been profound enough to allow us to be granted a patent on our discovery.

Now let's talk about some other forms of supplementary glutathione.

IV drips—Glutathione delivered, usually with a mix of other nutrients, through an intravenous line is 100 percent absorbable. The effects, however, only last from thirty minutes to three hours. That's because glutathione is in such high demand that our body uses it up readily. So, while it's hard to beat the ability of IVs to deliver glutathione right into the bloodstream, it isn't a practical way to address ongoing issues. But if you need a glutathione boost for an acute illness or to recover from a night of overindulging, IV therapy can be a good way to go. I'll talk more about it later in this chapter.

Oral supplements—The problem with oral glutathione supplements is that, once the active GSH enters the system, digestive enzymes in the gut dissolve and eat most of it up. It's not easy to tell how much glutathione actually gets absorbed but an educated estimate has it at about 15 percent at best, depending on how many milligrams were taken. You would have to take a large number of milligrams to have any results from taking oral supplements.

Sublingual (under the tongue) GSH drops—These oil-based drops are predicated on the idea that the glutathione will be absorbed through the mucous membranes and saliva glands in the mouth. Yet I've seen little evidence of such absorption, so I don't believe that the drops are particularly effective.

Topical GSH skin creams—Most makers of topical glutathione use liposomes to deliver the antioxidant through the skin. The best way to describe a liposome is to think of a microscopic sphere (or bubble) that the medicine or ingredient is incorporated into so it can be carried into the body through the oil (lipid) channels of the skin. However, these channels are slow and sticky, and thus not highly effective for delivering most molecules, including glutathione. And remember, glutathione is a very large and unstable molecule in its original form, making it especially hard to deliver. To date, I have not come across a liposomal glutathione that was effective. There are companies out there, too, that use different delivery systems for glutathione. I have not tried all of these skin creams and sprays, but again, because glutathione is a difficult molecule to get into the skin, I am skeptical about how well they work.

WHAT IS S-ACETYL GLUTATHIONE?

Supplements providing a particular nutrient or antioxidant often all get lumped together, when the truth is that they can be very different. For instance, one vitamin C tablet might have a different composition than another vitamin C tablet. And so it is with glutathione.

One form of glutathione on the supplement market is called S-acetyl glutathione, which can be found in tablet and topical forms. This is an acetylated form of glutathione, meaning that two carbon atoms, three hydrogen atoms, and one oxygen atom—collectively known as an acetyl group—have been added to affect how glutathione acts in the body.

In the medical field, some have stated that this newer form of GSH is better absorbed and more stable throughout the digestive tract than other forms on the market. And some laboratory data show that S-acetyl glutathione increases intracellular glutathione and has a positive effect on many oxidative stress biomarkers.

So, is this a good way for you to supplement your glutathione level? I have not put much time or research into S-acetyl glutathione, because I personally believe that the most effective glutathione molecule is one that's the closest to its natural form. It's called L-glutathione and it's the form of GSH that I have studied and prefer.

But I want to let you know about all the options available to you, so I would be remiss if I didn't highlight the potential of S-acetyl glutathione. Let's explore it further.

S-acetyl glutathione is a man-made molecule. Some evidence suggests that altering its form may improve supplementary glutathione's ability to remain intact in the gut, allowing a greater concentration to be absorbed into the bloodstream. If ongoing research continues to support this contention, then I believe it warrants asking another important question: Can S-acetyl glutathione penetrate the blood-brain barrier? If so, it might help address plaque in the hippocampus region of Alzheimer's patients. There may also be a use for S-acetyl glutathione in treating autoimmune diseases.

This is an area that I find rather intriguing but will need further research. In terms of using it as a supplement for your daily needs, right now I'd still advise using topical L-glutathione vs. oral S-acetyl glutathione as your go-to form of the antioxidant.

········•········

A CLOSER LOOK AT IV GLUTATHIONE

Glutathione is at work in our body 24/7, steadily and quietly tending to the numerous tasks at hand. We need it all day long. That's why you want to adopt the lifestyle habits that I outline in Action Plan One: It's what will help you produce a steady flow of GSH.

There can be times, though, when you need a large and quick infusion of the antioxidant, times when you are suffering from something acute, such as fatigue or the flu. When this is the case, there's a good likelihood that you will benefit from a more aggressive approach. That generally means seeking out intravenous (IV) glutathione and even other IV vitamins and solutions that allow you to take in high concentrations of nutrients at a rapid pace.

Intravenous glutathione hasn't been studied extensively, so most of the information about it is anecdotal. But it has been used for years by physicians, clinics, and wellness centers. My company supplied IV glutathione for over two decades, so naturally I am a proponent of these treatments, but I also feel that—and particularly if you have any disease that you're also taking medication for—you should discuss trying them with your medical professional before you make a decision. Also know that, although the FDA has neither approved nor banned IV glutathione, it is looking to make sure that all the material used in these treatments is injectable grade.

Intravenous glutathione has been in use for roughly the last twenty-five years. Sometimes it's called a glutathione "push," which refers to the administration of GSH at the end of any IV nutrition treatment. The first attempt at infusing fluids directly into the body was way back in 1492. It didn't work so well. More efforts followed until finally in 1832, an English physician was able to replace fluids in patients suffering from disease-related diarrhea. Fast-forward to the 1970s, when IV nutritional support became a common feature of hospital care.

Those early pioneers of IV therapy could hardly have foreseen what is happening now: IV wellness bars popping up all over, and especially in bigger cities, such as Los Angeles, New York City, Miami, Chicago, and Las Vegas. Las Vegas, in particular—who could forget the tagline "What happens in Vegas, stays in Vegas"?—has some of the highest numbers of IV nutrition clients. There are even several companies with mobile units.

Drinkers, hopeful of heading off a hangover, can call up and order IV gluta-thione hookups delivered to their hotel rooms. I've even heard of bachelor and bachelorette parties that include an IV nutrition staff.

Outside of alcohol recovery, IV glutathione use generally falls into these four general categories:

1. **To address overall wellness, fatigue, and/or chronic stress.** Peo-ple who live life at a busy, frenetic, stressful pace often seek out IV glutathione to help replenish their energy. So, too, do many health-conscious people who just want to ensure that everything in their body is operating at an optimal level.

2. **To remedy poor nutrition and digestion.** IV nutrition isn't a substi-tute for healthy eating, but it can help jump-start a change in behav-ior, and is a good way to nudge the digestive system back into good working order if you haven't been absorbing nutrients properly.

3. **To heal from illness or injury.** Glutathione's important role in im-munity and inflammation reduction can speed recovery.

4. **To enhance treatment for disease.** Many different medical condi-tions respond to glutathione and, although they may not be treated with GSH alone, the antioxidant can play a role in curing or lessen-ing symptoms of the disorders. For example, a number of Parkinson's patients have had great success treating the disease with high and frequent doses of intravenous glutathione (pushes), accompanied by other IV nutrients. The combination has been shown to greatly diminish the tremors associated with Parkinson's disease. Here are some other conditions that can be treated with IV glutathione:

   - Heavy metal detoxification
   - High triglycerides
   - Metabolic disorders, mainly type 2 diabetes
   - Viral infections
   - Tuberculosis
   - Chicken pox, shingles, herpes
   - Parkinson's disease

CAN YOU OVERDOSE ON GLUTATHIONE?

MOST OFTEN, THE ISSUE is not getting too much glutathione, but getting too little. GSH is an endogenous substance—we produce it ourselves and our body knows what to do with it. So, although glutathione overload has not been thoroughly studied, it's unlikely.

What's more probable than someone taking in or naturally generating too much glutathione is someone wasting their money. There's only so much GSH you can use within a given time. It's like pouring water into a glass that's already full. The water will just spill over the side without making its way into the glass. Likewise, if you load up on the antioxidant by, say, spraying a whole bottle on your body in one day, or having back-to-back IV treatments, some of that glutathione is going to go to waste.

In my twenty years' experience working with IV and topical glutathione, I have not seen one adverse side effect from excessive use. Outside of the occasional detox malaise or rash—effects that are not uncommon and relate to the release of toxins, not GSH—glutathione is remarkably safe.

Related questions that I get sometimes are: Can I build up a tolerance to glutathione? Will I need to keep increasing my doses in order to reach sufficient levels? And does this mean that if I am supplying my body with glutathione, it will stop making it on its own?

The answer to all these questions is NO!

You cannot build up a tolerance to glutathione supplementation. Your body needs it desperately on a daily basis and will always need it. But whether you're supplementing or not, your body will continue to make glutathione on its own. And if you stop supplementing, it will keep going regardless.

OTHER TYPES OF IV TREATMENTS

GLUTATHIONE TREATMENTS AREN'T THE only intravenous treatments you can find on therapeutic menus. One of the most popular, offered by many IV wellness centers, is something called the Myers Cocktail, a formula of intravenous vitamins and minerals that was pioneered by the late John Myers, MD. After Myers's death, many of his patients sought to continue treatments from another physician, Alan R. Gaby, MD. A renowned nutritional medicine expert, Gaby went on to refine the cocktail and document the results, presenting them at several medical conferences.

The Myers Cocktail includes magnesium, calcium, B vitamins (including B$_{12}$), and vitamin C, and is given by a slow infusion. This treatment allows the introduction of nutrients to the body in concentrations that cannot be achieved through oral administration. It's like the difference between using a hose vs. an eyedropper to water a plant—one gets the job done a lot more efficiently.

You many have noticed that I didn't list glutathione in the ingredients of the Myers Cocktail. That's because, traditionally, there is no GSH in the solution. Many medical professionals, however, add a glutathione push at the end of the treatment, a practice I heartily endorse. The glutathione goes in last because you want the other nutrients to do their jobs before the GSH enters your system and begins its role in taking care of cellular needs and detoxification.

So, when is a Myers Cocktail with a glutathione push called for? It's most effective for acute asthma attacks, migraines, fatigue (including chronic fatigue syndrome), muscle spasms, colds, chronic sinusitis, seasonal allergies, and chronic depression/anxiety.

Many wellness centers also offer their own variations on the Myers Cocktail and advertise them as treatments for specific conditions, such as migraines, food poisoning, flu, jet lag, hangover relief, or the more general hydrating, detoxifying, or beautifying. You can also get IV therapy with single nutrients, or if you're simply dehydrated, you can even just get a solution of electrolytes and saline. I call these IV "spot treatments."

One of the most common spot treatments is vitamin C. The reason for its popularity is simple: Vitamin C has always been the least expensive intravenous vitamin. This is not to discount its benefits, which although not as far-reaching or powerful as glutathione, are many. Vitamin C aids in many biological functions and is a vital antioxidant. Still, in high doses, vitamin C can go from being an antioxidant to a pro-oxidant, so you have to be careful about choosing an IV practitioner who knows what they're doing. Anyway, I believe that in the very near future you will see IV gluta-thione dethrone IV vitamin C as more people get wise to GSH. Everything that vitamin C can do, glutathione can do too—and it can do it better.

Before choosing a treatment that's right for you, I recom-mend (again) talking to your medical practitioner to help you make a selection. And, while perhaps I'm biased, I be-lieve that there's always room for a glutathione push, no matter what type of therapy you're having. All of us can use more GSH.

......●......

IV VS. TOPICAL GLUTATHIONE

When you're dealing with an issue where you need immediate relief, intra-venous nutrition can be a savior. Yet it has some drawbacks, which is why it's worth considering addressing your problem with topical glutathione or at least as a follow-up.

IV treatments can be expensive, running anywhere from $150 to $400, with a glutathione push running an additional $90 to $125. You'll have to do some research and perhaps travel to find the right place to have the ther-apy (though, as noted, some clinicians will come to you). Many people are not fond of having a needle placed in their arm, another downside. But the biggest drawback of IV glutathione is that, because GSH is in such high demand by the body, it's used up quickly. Most people use up an IV gluta-thione push in anywhere from thirty minutes to three hours.

Glutathione in a bottle can be applied at least every eight hours directly to the skin, at your convenience and at a lower cost. Many people have had success switching from IV glutathione to a self-administered solution. Let me share one man's story of making the switch.

Bob is an older gentleman who has several medical issues. He hap-pens to be a billionaire, so paying for IV treatments wasn't a problem for him, but he still wanted an easier treatment, since neuropathy had left him wheelchair-bound.

Neuropathy is a condition related to the peripheral nerves that causes numbness and/or weakness. Neuropathy is often a side effect of another medical condition, though it was unclear what was at the root of Bob's prob-lem. In any case, he was living with it, albeit without the ability to walk for nine years.

Anyway, here's the dramatic ending:

After ten days of using topical GSH (eight sprays a day), all the tingling in Bob's legs started to go away, and within thirty days, Bob was up and out of that wheelchair and walking with a cane. This type of Lazarus story can strain credibility—but I can vouch for its truly happening. Of course every individual is different, and everyone's root causes of disease and medical conditions are different. But my point here is that if Bob was out of a wheel-chair and walking in thirty days, there was a glutathione deficiency in there somewhere, and being able to treat himself at home proved just the ticket to clearing it up.

I have had other patients who also had success with at-home treatments of acute problems with glutathione. One of them, a registered nurse, had shin-gles, which was causing her considerable pain. Her doctor had prescribed antiviral pills for the condition, but they had led to heart palpitations that terrified her. As an RN, she was familiar with the benefits of IV glutathione, but when both a nutritionist she consulted and a pharmacist suggested she treat her shingles with a glutathione solution at home, she was skeptical. Yet she gave it a try anyway and was delighted when the painful burning from the sores decreased almost immediately. The sores began to heal rapidly too.

Next, she tried it on some fungus that had attacked her toenail; the fungus quickly went away.

Choosing a topical glutathione solution over IV glutathione is a personal choice, and that choice doesn't have to be the same in every instance. In some cases, an IV treatment might be the right call; in some cases, you may prefer a solution you can use at home. When you do opt for IV, however, be diligent about the place you choose. Research any prospective IV nutrition bar, and make sure it's accredited and medically supervised by an MD, or at least an RN, on-site at all times. And check the reviews. That will give you a fuller picture on which to base your selection.

14 DAYS TO KICK YOUR GSH-MAKING MACHINERY INTO GEAR

G OOD HEALTH DOESN'T HAPPEN OVERNIGHT. CAN IT HAPPEN IN TWO weeks? Of course, it depends on where you're starting from. If you have been subsisting on a diet of junk food, lying around on the couch, or have battled disease or other adverse health conditions, I have to be honest and say that you will likely need more than two weeks of a glutathione-producing plan to see a substantial difference. But here is what you can do in the short space of fourteen days: you can jump-start your GSH, making a very good start toward increasing glutathione throughout your body, which will allow you to better battle oxidative stress and its repercussions as well as increase your liver's ability to get rid of wastes and toxins.

The 14-day regimen involves following a vegetarian eating plan with easy recipes, adopting a supporting supplement plan, exercising moderately, and taking steps to reduce your stress levels. The effects of this combo of lifestyle practices will be subtle at first, but you will feel more energetic and your friends and family will notice the difference in you. Most important, they will give you a taste of what it feels like to be on a healthier path and provide a blueprint for habits that will keep your glutathione coffers full.

Remember the elderly women who were in good health and state of mind that researchers found also had high levels of glutathione (reacquaint yourself with them on page 96)? Keep these admirable ladies in mind going forward and know that, by taking control of your body's natural GSH-making machinery, you are setting yourself up to thrive well into your later years.

THE GLUTATHIONE DIET

In 2019, a study funded by the Bill & Melinda Gates Foundation was published in the prestigious medical journal *The Lancet*. Its alarming findings: In 2017, 11 million deaths across 195 countries were attributable to dietary risk factors. The leading risk factors were high intake of sodium (contributed to 3 million deaths), low intake of whole grains (another 3 million), and low intake of fruit (2 million). To sum it up, eating poorly will kill you, or at least lead to many years of illness.

Well, you might think, that's across 195 countries; in the US, we eat better than that. But consider this recent assessment in the *New York Times*: "Poor diet is the leading cause of mortality in the United States, causing more than half a million deaths per year. Just 10 dietary factors are estimated to cause nearly 1,000 deaths every day from heart disease, stroke, and diabetes alone. These conditions are dizzyingly expensive. Cardiovascular disease costs $351 billion annually in health-care spending and lost productivity, while diabetes costs $327 billion annually. The total economic cost of obesity is estimated at $1.72 trillion per year, or 9.3 percent of gross domestic product."

The 14-day glutathione eating plan is a model for eating well. It's designed to bolster your diet with wholesome, nourishing provisions while leaving out the unwholesome, damaging foods responsible for that roster of maladies listed by the *New York Times*. And, of course, the plan is going to help you raise your GSH levels. First and most significantly, by nixing such things as highly processed foods with preservatives and other chemicals that stress the body, it will help you "spend" less glutathione on fighting free radicals and dealing with heavy metals. Anytime glutathione is called to the front lines to deal with a food-related insult, it means there's going to be less of it available to keep your immune system and other physiological processes operating in full swing. As you may remember from Chapter 1, glutathione has the power to regenerate itself; however, this is not an infinite cycle. You can only take your favorite boots to the shoe repair shop so many times before they wear out beyond rehabilitation. With GSH as well, something eventually has to give. So, what you don't eat during these fourteen days is just as important, perhaps even more so, than what you do eat.

The second way the diet impacts GSH levels is by providing you with foods that contain glutathione or other substances that promote the production of glutathione. Some antioxidants can be readily obtained from food (in fact, some, such as vitamin C, can *only* be acquired from food). Unfortunately, the same cannot be said for glutathione; however, the body does absorb GSH as well as its building blocks (most notably cysteine) through the digestive system. To maximize the glutathione available to you, this 14-day plan is made up of *fresh* foods: GSH content is substantially reduced in foods that have been processed or preserved.

Glutathione, of course, is not the end-all, be-all of good health. It can't do the job alone, which is why it's so important that your diet be not only GSH-boosting, but extremely nutritious. This eating plan is designed accordingly: It's well balanced, providing enough protein, carbohydrates, and fat to help your body run at its optimal level. The meals are built largely on fruits and vegetables, whole grains, vegetarian proteins, and healthy fats. It has very little sugar, and what sugar it does have is natural. Packing your diet with as many wholesome foods as possible may be especially important now. Research is showing that climate change is placing more carbon dioxide in the air and it's affecting the nutrient makeup of foods. Anticipating that there will be even more CO_2 in the air to come, USDA scientists exposed rice crops in Japan and China to high levels of the gas and found that the crops had significantly less protein, iron, zinc, and B vitamins than regular rice crops. It's an unsettling harbinger of the future and may even be taking hold now.

I have personally been consuming The Glutathione Diet for twenty years. My preference is to maintain an organic, primarily vegetarian diet, something you may not ultimately choose to do going forward. But in the interest of kick-starting your glutathione-making machinery, I'm asking you to eat mostly vegetarian (with a little fish thrown in) for just two weeks and to choose organic foods as often as possible.

When you look at the chart that identifies foods with the highest con-centrations of glutathione and cysteine, you will likely notice that meat and poultry are fairly rich in both. Why, then, do I suggest you refrain from eat-ing all types of meat during the initial 14-day kick-start diet? Let me explain.

When you eat meat, the toxins that the animal has ingested end up in your intestines, which can create oxidative stress in your body and tax your glutathione system. (Fish can have toxins, too, which is why I recommend ONLY those with the lowest levels—see page 148.) The whole goal of rais-ing your glutathione levels is to help your liver get rid of toxins and to clear your detoxification pathways. Eating meat may bring in more glutathione, but it also makes you use more glutathione, defeating the purpose and slow-ing the systemic detoxification process. Ultimately, it won't help and will likely hinder your body's ability to raise its GSH levels.

Eating meat, of course, is a personal choice, but I do want to add one more caution. If, when you eat meat, your stools are hard and not easily eliminated by your body, it's a sign that you're not digesting it properly (or drinking enough water). But if you are a meat lover, and it's not something you want to completely cut out, I suggest picking two or three days a week to indulge in your favorite types of meat protein, then go vegetarian (or at least pescatarian) the rest of the week. And I highly urge you to choose organic, vegetarian-diet-fed types of lean meats, and go light on red meat in particular.

During the 14-day diet, I'll also ask you to refrain from foods not on the diet, even if they are rich in glutathione or cysteine. There's a reason for that. Take oranges, for example. Oranges are very acidic and can lower the body's pH levels to a point that interferes with detoxification. (Keeping the pH at normal to high helps with raising GSH levels.) However, once you get through the initial two weeks, feel free to work the occasional citrus fruit into your diet.

Despite these restrictions, The Glutathione Diet is very diverse and easy to follow. You'll find that many of the dishes repeat (and week two is a repeat of week one). I find that most people are creatures of habit when it comes to breakfast and lunch, but like variety in the evening; the plan is designed accordingly. There is some cooking involved (and blending—make sure you have a blender on hand), but the recipes are simple and require no fancy ingredients or utensils. As with any prescribed eating regimen, a little advanced planning is needed, but nothing that you can't take in stride. Think of it as your entrée to a new way of eating and an introduction to recipes that will serve you well over time.

FAQ

I'm betting you have a few questions. Here are the answers.

Q. Without meat, and with fish only once in a while, will I get enough protein?

A. The amount of protein we need to stay healthy has often been overstated. The recommended daily amount for protein is 0.8 gram for every kilogram of body weight (or as we know it, 2.2 pounds). So, if you are a 130-pound woman, that means you need about 47 grams of protein per day. A 170-pound man would need about 63 grams. If you're very athletic, you may need slightly more: 1.2 grams per kilogram of body weight.

We get protein from a lot of sources (even broccoli and avocado contain protein), including many of the foods included in my meal plan. The morning shake alone has about 17 g protein, and you'll pick up more grams throughout the day as you eat nuts, whole grains, and certain vegetables. Remember, too, that no day has to be perfect. You may eat less protein on one day and more on another; what's important is the overall balance.

Q. Why isn't The Glutathione Diet a keto diet?

A. There are a lot of reasons the ketogenic (keto) diet is popular right now, among them the fact that it's helped many people lose weight without feeling hungry. Like the paleo and Atkins diets, the keto diet is low in carbohydrates, but unlike those protein-centric eating regimens, the keto diet focuses on fat—sometimes calling for as much as 90 percent of total calories. The keto diet has been found to reduce epileptic seizures in children and adolescents, which has led some researchers to ask why. One possible reason: the keto diet increases glutathione production, at least in lab rats and mice. In particular, the diet seems to increase antioxidant activity in the mitochondria, which may have an anticonvulsant effect.

So the question of why The Glutathione Diet isn't keto is a good one. Here's my thinking: There are a lot of ways to increase glutathione and I believe that the best way is to eat wholesome GSH-enhancing foods. The Glutathione Diet is relatively low in carbohydrates and has a fair amount of fat, but it's much more moderate than the keto diet, which may ultimately be hard to maintain—and where will that leave you? Better, I believe, to eat in ways that are sustainable (both in terms of helping you stick to it and in terms of its effect on the planet), delicious, and varied. Keto diets can also be high in saturated fat (though it's also possible to eat in a vegetarian keto way), and that may be a risk factor for heart disease.

If you're already on a keto diet and feel it's working for you, by all means stay on it. But do give this Action Plan a read to see what GSH-boosting foods you might be able to work into your current eating regimen.

Q. Does this diet work with intermittent fasting?

A. Intermittent fasting (IF), a diet that involves alternating between periods of eating and fasting, is more about timing than it is about the foods you eat. In that sense, it can be combined with glutathione-boosting foods, especially if you are trying to detox (see page 169 for more about IF and detoxing). That said, I would suggest keeping a traditional schedule during the two weeks you spend on The Glutathione Diet; that way you can focus on eating healthfully and incorporating some very specific foods into your eating regimen. To my mind, that's enough change for someone just starting out. Once you're past the two weeks, feel free to experiment with IF.

Q. What should I drink?

A. Water and lots of it. I recommend eight glasses a day. To many people, that's boring, but I stand firm on the idea that it will help all of your body's functions, including the manufacture of glutathione. So, make water, plain water, not carbonated, your first choice.

To my knowledge, neither tea nor coffee depletes glutathione, so you can also drink both. Just keep it in moderation and don't let cups of either replace your daily glasses of water. If you're keeping your coffee/tea intake within reason, adding a little milk or cream won't hurt. For sweeteners, refer to the list on page 140.

Coconut water can be used to replace regular water; so can hot water with a squeeze of lemon and a little honey (that's my personal favorite) or herbal tea. Stay away from sugary (or artificially sweetened) sodas and juice. Kombucha is okay to drink once in a while, but if you're a fan, give it a rest until after the two weeks are up. Most certainly, abstain from drinking alcohol. When you drink alcohol, your need for glutathione goes up to the nth degree, negating the GSH-boosting benefits of the diet.

Q. I work out a lot. Can I eat more than the diet calls for?

A. Yes, but only supplement the diet with foods that will help raise your glutathione levels. See the list of snacks on page 152.

A FEW MORE NOTES ON THE DIET . . .

Healthy fats. Indians have long used clarified butter (we call it ghee) and its health benefits are plentiful. Clarified butter is butter with the solids removed and it's rich in the antioxidant vitamins A and E (another molecule that glutathione depends on to keep functioning properly) as well as vitamin D. Butter, including clarified butter, is a saturated fat and saturated fats can raise cholesterol and triglyceride levels. That said, the amount used in this diet is low and, as noted, clarified butter has other benefits. I believe that using it in moderation is safe, particularly since The Glutathione Diet doesn't include other sources of saturated fat such as dairy products (whey protein is dairy-derived but has miniscule amounts of saturated fat). The same goes for coconut oil, which is called for here sparingly. Coconut oil is also a saturated fat; however, the jury is still out on whether it raises risk factors for heart disease. Because we still don't know, use it in moderation.

One fat we know is healthy is extra-virgin olive oil because it's monosaturated, a type of fat that doesn't raise cholesterol or triglycerides. It's called for often here and should be your go-to fat. Oils to avoid: soybean, corn, safflower, and canola. These oils are rich in omega-6s, fatty acids that can promote inflammation.

MCT and avocado oils. These are two other healthy fats I want to mention. Although the recipes for this diet primarily call for olive or coconut oil, I also like to cook with MCT and avocado oils. MCT oil refers to medium-chain triglyceride fat and is usually derived from coconut oil (regular coconut oil is not a medium-chain triglyceride fat). The mitochondria in our cells love to feast on MCT and avocado oil, assisting in energy production. You can use them in place of any of the fats called for in this diet. You might also put a few teaspoons in your morning shake. MCT oil is available at such places as Whole Foods, Walmart, and GNC, and online.

Sweeteners. If you must use a sweetener in your shake, coffee, or tea, I recommend using one of these four choices: Stevia, agave syrup, monk fruit extract, or coconut nectar (palm) sugar. These all have very low glycemic levels (meaning they don't rapidly raise blood sugar) but are as sweet as sugar. You can get these at Whole Foods, some natural food stores, and online.

Don't get too nutty about Brazil nuts. Brazil nuts are the best source of GSH-raising selenium, and a prime player in this diet. However, because they are so rich in selenium, use caution—the trace element can be toxic if you consume too much of it. Selenium toxicity can cause anything from gastrointestinal distress to heart, lung, and kidney failure. So, use caution. Limit yourself to four Brazil nuts a day.

THREE GSH-PROMOTING NUTRIENTS

WHEN YOU THINK ABOUT glutathione in the diet, don't just think about, well, glutathione in the diet—that is, include foods that stimulate glutathione production, not just foods that contain the antioxidant. Here are three of the most important nutrients to include in your daily diet.

1. Nondenatured whey isolate protein. Whey protein is a derivative of milk, and it contains all the key amino acids for glutathione production (cysteine, glycine, and glutamate) as well as a unique cysteine compound that easily converts to glutathione. Whey isolate protein is available in natural food stores and many supermarkets.

I am not a big fan of dairy, but this is an exception—with a few warnings. The whey isolate protein MUST be bioactive, meaning it contains living organisms. Choose nonpasteurized and non–industrially produced whey that has no pesticides, hormones, or antibiotics. The reason I recommend nondenatured (sometimes called undenatured) whey is that the denaturing process sometimes used to produce whey can diminish the amino acids' biological action. ("Denaturing" refers to the breakdown of the normal protein structure.) In other words, you'll be wasting your money on protein that doesn't provide the nutrients you need.

If you are allergic to milk or lactose, you won't be able to use whey protein. In that case, I recommend using a vegan, plant-based protein powder in your smoothies instead. Whey isolate protein has less lactate than whey protein concentrate, but the only way to tell if you can handle it is to try it. Or just skip it altogether and go with the vegan protein powder.

2. Selenium. This trace mineral, an antioxidant itself, is a glutathione cofactor, meaning it's a substance needed for glutathione activity. Research has also shown that it increases the production of certain types of glutathione molecules. The Recommended Dietary Allowance (RDA) for selenium for adults is 55 mcg, and (way) more isn't necessarily better. Ingest no more than 400 mcg, at which point selenium can become toxic.

Several of the foods in The Glutathione Diet are rich in selenium, especially Brazil nuts—the best source. Brazil nuts, in fact, are one of my secret weapons. I keep snack bags of these in my car to snack on both to quell cravings for crunch and to keep my energy level up (this is, though, a secret weapon I eat in moderation—see page 140).

On page 145, I've listed the top selenium choices and the amounts of the mineral they offer. These are my top choices, though there are also some other sources that are rich in selenium, including cottage cheese, tuna, halibut, sardines, shrimp, ham, steak, turkey, and chicken. Still, I urge you to choose the sources on the list as your go-to selenium foods.

The other sources can be problematic. In particular, tuna can contain mercury, ham may have nitrates, and these other animal products also have the potential to contain heavy metals, antibiotics, hormones, or other contaminants. If you do opt for some of the selenium sources on this second list, at least look for products that are organic or otherwise produced responsibly.

Bear in mind, too, that you will likely get small amounts of selenium in other foods you eat throughout the day, including those on The Glutathione Diet. Those bits and pieces will add up, so be sure to eat onions, spinach, garlic, yeast in bread, nutritional yeast, lentils, raisins, walnuts, and cashews.

3. Vitamin C. Research has shown that vitamin C increases the glutathione in red blood cells. One study, conducted at Arizona State University, found that supplementing with 500 mg vitamin C per day increased glutathione levels by 50 percent! See chart, page 144, for your best food sources.

········●········

FOODS WITH THE HIGHEST CONCENTRATIONS OF GLUTATHIONE

The body can both absorb already-made glutathione from food and make glutathione from its building blocks. Cysteine is the most difficult to get of those three building blocks, so along with the glutathione-rich foods listed here, also eat plenty of foods that are good sources of the amino acid.

FRUIT

Avocado (about ¾ fruit) 20.6 mg

Strawberries (about ¾ cup, halved) 6.9 mg

Grapefruit (about 6 sections) 6.5 mg

Cantaloupe (about 1 large wedge) 6.1 mg

Papaya (about 1 cup, 1-inch pieces) 5.8 mg

Watermelon (about ¾ cup, diced) 5 mg

Peaches (about ¾ cup, sliced) 5 mg

Oranges (about 6 sections) 4.8 mg

VEGETABLES

Asparagus (about 5 large spears) 21.8 mg

Spinach (raw) (about 3 cups) 11.4 mg

Okra (8–9 pods) 11.3 mg

Winter squash (about 1 scant cup, cubed) 11 mg

Potatoes (with skin) (about ½ medium) 11 mg

Tomatoes (about 3 large, thick slices) 7.5 mg

Carrots (about 2 small) 5.9 mg

Spinach (cooked) (about 2 cups) 5.7 mg

MEAT, POULTRY, FISH

Pork chop (about 3½ ounces) 18.9 mg

Steak (about 3½ ounces) 12.3 mg

Hamburger (about 3½ ounces) 11.8 mg

Chicken (about 3½ ounces) 7.7 mg

Fish (cod) (about 4 ounces) 5.7 mg

FOODS WITH THE HIGHEST CONCENTRATIONS OF CYSTEINE (PER 100 GRAMS)

FRUIT

Orange juice (about 3½ ounces) 18 mg

Bananas (1 small) 17 mg

Oranges (about 6 sections) 10 mg

Tangerines (about 14 sections) 10 mg

Peaches (about ¾ cup, sliced) 6 mg

VEGETABLES

Beans (pinto, black-eyed peas) (about ¾ cup, cooked) 50 mg

Spinach (about 3 cups, raw; 2 cups cooked) 35 mg

Peas (⅔ cup, cooked) 26 mg

Potatoes (skin on) (about ½ medium) 24 mg

Cauliflower (about ¾ cup, cooked) 22 mg

Broccoli (about ¾ cups, cooked) 21 mg

MEAT, POULTRY, FISH

Liver (about 3½ ounces) 410 mg
Chicken (about 3½ ounces) 358 mg
Pork chop (about 3½ ounces) 350 mg
Steak (about 3½ ounces) 279 mg
Hamburger (about 3½ ounces) 268 mg
Cod (about 4 ounces) 259 mg
Pollock (about 4 ounces) 227 mg

MISCELLANEOUS

Walnuts (1 cup) 345 mg
Peanut butter (about 6 tablespoons) 315 mg
Bran flakes (about 2¼ cups) 229 mg
Eggs (about 3 large) 278 mg

Sources: Dean P. Jones, Ralph J. Coates, Elaine W. Flagg, John W. Eley, Gladys Block, Raymond S. Greenberg, Elaine W. Gunter, and Bethany Jackson, "Glutathione in Foods Listed in the National Cancer Institute's Health Habits and History Food Frequency Questionnaire," *Nutrition and Cancer* 17, no. 1 (1992): 57–75; USDA FoodData Central.

BEST SOURCES OF OTHER GLUTATHIONE-BOOSTING NUTRIENTS

VITAMIN C

Red pepper (1 cup, chopped) 190 mg
Green pepper (1 cup, chopped) 120 mg
Broccoli (1 cup, chopped, cooked) 101 mg
Brussels sprouts (1 cup, cooked) 97 mg
Strawberries (1 cup, halved) 89 mg
Grapefruit (1 medium) 88 mg
Papaya (1 cup, cubed) 85 mg
Orange (1 medium) 70 mg
Cantaloupe (1 large wedge) 37 mg
Tangerine (1 medium) 23 mg

SELENIUM
Brazil nuts (¼ cup) 637 mcg
Brown rice (1 cup, cooked) 11 mcg
Egg (1 large) 18 mcg
Wheat germ (2 tablespoons) 9 mcg
Sunflower seeds (¼ cup) 18 mcg
Pinto beans (1 cup, canned) 9 mcg

WHAT'S NOT IN THE GLUTATHIONE DIET, AND WHY

What you don't eat is as important as what you do eat during the fourteen days you'll spend kick-starting your GSH production. What's more, you'll be following The Glutathione Diet for only two weeks—what happens when those fourteen days are over? Hopefully, you will continue to eat in very much the same way as laid out by the plan and including many of the foods high in glutathione and glutathione-building blocks in your weekly menu. But going forward, it's also important to be aware of the foods that sap GSH and, frankly, are just unhealthy in general. With that in mind, here is a rundown of the foods you should bypass in your grocery store.

THE DIRTY DOZEN

Chemicals used in farming—pesticides, herbicides, fertilizers, antibiotics—make their way into our food supply, doing harm in various ways. One of those ways is by diminishing glutathione. The more toxins you take in, the more you'll need to spend from your glutathione "bank account" to get rid of them. If you are ingesting a lot of chemicals, you're going to use up your GSH supply, leaving little to none available to fight aging and disease.

In a perfect world, we would all be able to eat certified organic or, at the very least, foods that are produced without pesticides and antibiotics. In the imperfect world we live in, that's not always possible. If you have to pick and choose among healthfully farmed food, I recommend that you try to avoid what's known as the Dirty Dozen, a list of twelve fruits and vegetables that the Environmental Working Group updates every year (ewg.org). The ones on this list are the twelve the group's 2019 study found contain high levels of pesticide residues. Here's the list (starting with the dirtiest):

- Strawberries
- Spinach
- Kale
- Nectarines
- Apples
- Grapes
- Peaches
- Cherries
- Pears
- Tomatoes
- Celery
- Potatoes

THE CLEAN FIFTEEN

The Environmental Working Group also has a list of fruits and vegetables that you can feel good about buying even when they're not organic. The clean fifteen (starting with the cleanest) are:

- Avocados
- Sweet corn
- Pineapples
- Frozen sweet peas
- Onions
- Papayas
- Eggplants
- Asparagus
- Kiwis
- Cabbage
- Cauliflower
- Cantaloupes
- Broccoli
- Mushrooms
- Honeydew melon

PROCESSED FOODS

Processed foods is a term that gets thrown around a lot, but what does it really mean? Is bread a processed food? What about cereal and yogurt? Granted, you don't pick any of these off a vine (even plain yogurt involves some mixing and manipulating), but to my mind, "processed foods" refers to foods that have been combined with chemicals designed to help preserve them, color them, texturize, and/or artificially flavor them. Processed foods also tend to be saturated with sugar (in all its variations from white sugar to high-fructose corn syrup), unhealthy fats, and salt.

Processed foods have an effect on glutathione levels that's similar to the effect of foods carrying chemical residues: Both overwork the body's toxic waste removal system, meaning that they use up the glutathione you need to stay at your healthiest. Eating a lot of processed foods can also lead to inflammation and disease. This is where glutathione can help but—catch-22—not if your GSH stores are substantially lowered because you've been living on processed foods. Something else to consider is that a junky diet is inherently low in vitamins, minerals, antioxidants, and other nutrients that are glutathione's teammates. To keep the free radical–fighting, detoxifying team intact, choose whole, minimally processed foods as often as possible.

The way to tell whether a food is highly processed is to look at the information on the label. If the ingredients list reads like a chemistry professor's version of *War and Peace*—e.g., chromium amino acid chelate, sodium selenite phylloquinone, polydextrose, sucralose, dipotassium phosphate, and so on—it's a good indication that you should forgo this particular food. What you want instead are very short ingredients lists, comprised of names that you recognize. In other words, ones that read more like: whole wheat flour, water, yeast, sea salt, sunflower seeds. Remember, too, that ingredients are ordered by weight. The whole-grain bread that I just described is mostly made up of whole wheat flour and water.

FOODS CONTAMINATED WITH HEAVY METALS

Even the Food and Drug Administration (FDA), the most conservative of agencies, acknowledges that such metals as arsenic, lead, cadmium, and mercury are in our food supply. Because many factors are involved, including where food is grown and the water used for growing it, it's difficult to know exactly what the heavy metal levels are in the foods you buy at the supermarket. Many organizations, however, have done metals testing in order to make health recommendations. Here's what we know right now.

Seafood. The heavy-metal content of seafood has been heavily documented. Mercury is the main contaminant in fish, particularly big fish, which eat the little fish and all the mercury they contain too. On the National Resources Defense Council's fish-to-avoid list:

- King mackerel
- Marlin
- Orange roughy
- Shark
- Swordfish
- Tilefish
- Ahi tuna
- Bigeye tuna

Also avoid farmed fish—seafood and freshwater fish that are raised in controlled pens. These fish can be contaminated with toxins and the farming causes environmental pollution. Farmed salmon, in particular, has been found to harbor high levels of toxins. It's a fatty fish and heavy metals and chemicals accumulate in fat tissue.

Just as there is a safe food list to counter the Dirty Dozen fruits and vegetables, there is also a roster of "green" seafood with low mercury levels. According to the Monterey Bay Aquarium's Seafood Watch, the list includes:

- Atlantic mackerel
- Freshwater coho salmon
- Pacific sardines
- Wild-caught salmon from Alaska, both fresh and canned
- Pole- and trawl-caught albacore tuna and sablefish (black cod)

If you'd like to keep it simple, you might stick to Dr. Mark Hyman's easy-to-remember SMASH guideline: Eat only sardines, mackerel, anchovies, wild salmon, and herring.

Rice. Rice? Yes, rice. I'm not saying never eat rice—I include it in The Glutathione Diet—but here is a heads up. A few years ago, data from the FDA indicated that rice had surprisingly high levels of arsenic. It may, however, depend on where your rice comes from. *Consumer Reports* did tests and found that white basmati rice from California, India, and Pakistan, and sushi rice from the US, on average, have half the arsenic of most other types of rice. Other types of US-grown (but not California) rice (e.g., long-grain) had higher levels of arsenic. Bear in mind, too, that brown rice has higher levels of arsenic than white rice. Although it may have more health-promoting fiber, the heavy metal accumulates in the outer bran layer of brown rice. If you're going to choose brown rice, try to buy organic varieties from California, India, and Pakistan. Also think about other whole-grain alternatives, such as:

- Millet
- Spelt
- Farro
- Whole wheat couscous
- Nongrain alternatives, such as cauliflower rice (Many stores now sell "prericed" cauliflower so you don't have to go through the trouble of running it through the food processor yourself.)

Bone broth. Bone broth is probably one of the oldest foods around, and it's been rediscovered of late, prized for promoting good digestion. Except . . . in 2013, researchers in the UK looked at the heavy metal content of bone broth and found that, because metals tend to accumulate in the bones, the chicken bone broth they tested had high levels of lead. To be fair, a later study did not find that bone broths had metals in toxic amounts. Still, it's something to be aware of so, if you do enjoy these types of soups, shop carefully. Some natural food markets sell organic versions. Add a ton of veggies and some organic chicken or turkey and you have a comforting and nutritious meal.

Artificial sweeteners. I've just told you to stay away from foods saturated with sugar. Don't, though, take that as an indication that you should switch to food and drink flavored with artificial sweeteners. For the most part, these alternative sweeteners are chemicals and your liver, with a GSH assist, will have to work hard to get rid of them. Here is a list of the "do not consume" sweeteners with their brand names in parentheses:

- Acesulfame potassium (Sunnett, Sweet One)
- Aspartame (Nutrasweet, Equal)
- Saccharin (Sweet 'N Low, Sweet Twin, Sugar Twin)
- Sucralose (Splenda)

Alcohol. The Glutathione Diet asks you to be a teetotaler for fourteen days. Do you have to abstain from alcohol always? No. I myself don't drink alcohol very often, a personal choice. It may not be your preference, but I want to just make you aware that, as noted in Chapter 4, it's not just that your favorite cocktails may have some harsh effects, including prompting dehydration. Alcohol also acts as a solvent, dissolving toxins that might otherwise not be able to make it into the bloodstream. Alcohol opens the gate for these harmful substances, which will eventually end up in the liver—where they will need extra glutathione to get rid of them. Even one glass of wine can deplete your glutathione supply. Keep this in mind the next time someone offers you a gin and tonic!

7-DAY TEMPLATE

	DAY 1	DAY 2	DAY 3
Breakfast	Get-the-Glow Protein Shake	Get-the-Glow Protein Shake	Get-the-Glow Protein Shake
Midmorning Snack	¼ cup Nutty Raw Mix	¼ cup Nutty Raw Mix	¼ cup Nutty Raw Mix
Lunch	1 Super Fruit Bowl Field of Greens Salad 1 date	1 Super Fruit Bowl Hearty Avo-Veg Sandwich 1 date	1 Super Fruit Bowl Field of Greens Salad 1 date
Dinner	Ginger-Garlic Vegetable Stir-Fry over Rice	Garlic and Thyme Roasted Vegetables	Savory Mixed Bean Soup

TWO WEEKS OF GLUTATHIONE-BOOSTING MEALS

All of the recipes for the 14-day plan follow in Appendix B. A few final words before you get started:

I've been practicing this glutathione lifestyle for a very long time; it isn't always easy. So, don't be too hard on yourself. Just do the best you can each and every day. I find I'm most successful when I have a positive perspective. In that vein, think of this plan as a gateway to feeling better than you probably have in a long time. Focus on how a small amount of discipline offers you health benefits that far outweigh the steps it takes to get there. A lot of delicious foods are available to you on this plan, so have fun with it! A few helpful hints for upping the fun factor:

- Pick a favorite friend, family member, or work buddy to do the 14-day Glutathione Diet with you.
- Encourage each other. Go on brisk walks together and while doing so discuss victories and challenges.
- Get a journal to keep track of your progress. Keep dated notes of your feelings and questions.
- Be extra nice to yourself during this process, and remember, you're worth it!

DAY 4	DAY 5	DAY 6	DAY 7
Get-the-Glow Protein Shake	Get-the-Glow Protein Shake	Get-the-Glow Protein Shake	Get-the-Glow Protein Shake
¼ cup Nutty Raw Mix	¼ cup Nutty Raw Mix	¼ cup Nutty Raw Mix	¼ cup Nutty Raw Mix
1 Super Fruit Bowl Hearty Avo-Veg Sandwich 1 date	1 Super Fruit Bowl Field of Greens Salad 1 date	1 Super Fruit Bowl Hearty Avo-Veg Sandwich 1 date	1 Super Fruit Bowl Field of Greens Salad 1 date
Steamed Vegetables with Guacamole	Stir-Fried Vegetables with Quinoa	Lentil and Vegetable Soup	Pan-Seared Fish with Raspberry Vinaigrette

Days 8–14: Repeat Days 1 through 7

WHEN YOU JUST NEED A LITTLE SOMETHING . . .

The midmorning snack in The Glutathione Diet eating plan will probably hold you over well until lunchtime. Nuts, because of their protein and fat content, are very satisfying. If you find yourself feeling hungry after exercise, or at any point during the day, feel free to snack. The only time I urge you not to snack is in the later part of the evening, when you're likely to be sedentary and have little time to digest before bed. The one exception: raw or leftover cooked vegetables, which are always an option. For the rest of the time, here are some healthy and satisfying choices:

- **Hard-boiled egg:** Eat the egg the day you make it; eggs can be a petri dish for bacteria.
- **Tablespoon or two of almond, peanut, sunflower seed, or cashew butter:** Buy the unsweetened and organic kinds.
- **Small handful of roasted-in-the-shell pistachio nuts:** Flavored versions (e.g., garlic, onion, jalapeño) are okay.
- **1 ounce dried root vegetable chips (e.g., Terra):** Look for ones made with healthy oils.
- **1 cup homemade kale chips:** The kind in packages can be heavily salted and expensive (for what you get).
- **1 cup cucumber cubes:** Flavor with a little salt and dried red pepper flakes if desired.
- **Small bowl of berries:** A great way to get more antioxidants into your day.
- **1 cup sugar snap peas:** I won't say these are as addictive as potato chips, but the crunch makes these vegetables particularly satisfying.
- **Small box of raisins:** This will up your selenium intake a bit. However, raisins contain a lot of sugar, so keep it to a small box or the equivalent (1.5 ounces).

- **1/2 cup roasted chickpeas:** To make roasted chickpeas, drain a can of chickpeas, pat dry, then toss with a little olive oil and salt. Roast in a 400°F oven for about 20 minutes. Sprinkle with the seasoning of your choice while they're still warm. Store in an airtight container in the refrigerator.

SUPPORTING SUPPLEMENT PLAN

Note: Please check with your health-care provider before starting any supplement regimen.

To reiterate something I've been saying throughout this book, glutathione works in partnership with other nutrients to keep the body vibrant and healthy. Eating a nutritious diet goes a long way toward ensuring that you have adequate amounts of those partners, but considering the world we live in today, it's not always enough. As noted earlier, the USDA has found that increased carbon dioxide in our air is reducing the nutrient value of our food supply. Several studies have also shown that fruits and vegetables are now lower in nutrients than they were in previous years, due to soil depletion and the differences in plant breeding (the more resistant to pests, it seems, the lower the nutrient levels of a plant).

One study, done by researchers at the University of Texas at Austin, compared USDA agricultural data from 1950 and 1999, and found remarkable differences in vitamin C, B vitamins, calcium, iron, and protein. It was published in 2004; imagine what adding another decade and a half of changes may have done to the food now landing in our supermarkets. Because of these changes, I add supplements to my daily routine, and recommend that you do too. Here is a sample daily regimen that will help boost both your glutathione levels and levels of other crucial nutrients.

DAILY SUPPLEMENT PLAN

TAKE IN THE MORNING	TAKE IN THE EVENING	TAKE WITH A LARGE MEAL (SUCH AS DINNER)
300 mg coenzyme Q10 100 mcg selenium *Note: Only take selenium if you're not eating Brazil nuts* 1,000 mg vitamin C	600 mg magnesium glycinate *Note: Start with 100 mg and slowly increase to 600 mg if tolerated and no diarrhea* 400 mg mixed tocopherols vitamin E *Note: I recommend a brand called Unique E, but whatever brand you choose, look for "mixed tocopherols." Vitamin E activity comes from not one but a family of related molecules called tocopherols and tocotrienols. Tocopherols are likely involved in strengthening glutathione activity.*	650 mg betaine HCl + 1 capsule of digestive enzyme *Note: Both of these help the body break down food into the various components we need to function. Betaine is similar to stomach acid. Natural stomach acid decreases as we age, making it hard to digest food and often leading to indigestion.*

A BALANCED EXERCISE ROUTINE

Exercise, for all its health benefits, can create free radicals that tax the antioxidant defense system. But we also know that exercise can increase glutathione levels—at least moderate exercise can (see Chapter 6). I realize, though, that keeping exercise moderate doesn't work for everybody. Some of you may be amateur or even professional competitive athletes; others may simply enjoy intense workouts.

Not to worry, but you may need to take a few extra steps. For one thing, if you work out hard and often, have your GSH level tested. That will let you know what you're working with. If your level is low, it's a good idea to bump up the amounts of glutathione-boosting foods and even supplement with glutathione solution daily. Taking the supporting supplements mentioned above can help too. A once monthly IV treatment with a glutathione push is an option for keeping your GSH level up as well.

The workout plan that follows is a one-week schedule (you repeat it the second week) that fits in with the 14-day GSH kick-starter plan. The schedule establishes guidelines for exercising; you will need to tweak it according to your own abilities and tastes. You may like spinning for cardio; another reader may prefer brisk walking. Someone with access to a gym might like the elliptical trainer. The great thing about aerobic exercise is that there are so many ways to do it.

I hope you will also use these guidelines to help you continue being active beyond the initial fourteen days. The plan provides all-around fitness, hitting the cardiovascular system, muscle strength, and flexibility. Flexibility, which is often forgotten in exercise regimens, is of paramount importance as you age as it will help you stay agile. And don't be discouraged if it feels futile at first. It took me until I was forty-six, but I can finally touch my toes!

GLUTATHIONE-CONSERVING WORKOUT SCHEDULE

Sunday
ACTIVITY: Leisurely walking
DURATION: 30 to 60 minutes

Monday
ACTIVITY: Cardio workout of your choice with intervals
EXAMPLE: Jogging or using the elliptical machine at a moderate* pace for 3 minutes followed by jogging at a vigorous** pace for 1 minute; repeat
DURATION: 30 minutes (or as long as you can, working toward 30 minutes)

---

* Note: **Moderate** means fast-paced enough to get your heart rate up, but not so fast that you can't sustain it for 3 minutes. What this means for you will be individual, but by most exercise scales, moderate feels like you can keep going for an indefinite time and, while you may be breathing harder, it's not so hard that you can't carry on a conversation.

** **Vigorous** means quick enough to really tax your system, but not so fast you can't sustain it for 1 minute. Again, this will be individual, but it's often defined as on the verge of becoming uncomfortable and hard enough that you are somewhat short of breath.

Tuesday

ACTIVITY: Strength training using your own body weight such as sit-ups, push-ups, squats
DURATION: As long as it takes to do 2 to 4 sets of 10 repetitions, depending on your capabilities

Wednesday

ACTIVITY: Yoga routine or all-over body stretching routine
EXAMPLE: Find an online class that matches your ability.
DURATION: 30 to 60 minutes

Thursday

ACTIVITY: Cardio workout of your choice with intervals
DURATION: 30 minutes

Friday

ACTIVITY: Strength training using your own body weight
DURATION: As long as it takes to do 2 to 4 sets of 10 repetitions, depending on your capabilities

Saturday

ACTIVITY: Yoga routine or stretching exercises
EXAMPLE: Find an online class that matches your ability.
DURATION: 30 to 60 minutes

PROTECTION FROM THE SUN

Sun and heat promote free radical production and, of course, you know by now that exposure to UV rays promotes skin cancer and accelerates aging. So, what can you do during these two weeks to protect yourself?

First and foremost, stay in the shade whenever possible. When you cannot, wear protective clothing, including a hat. This is not always practical. I live in Southern California, so I'm aware that most people aren't fully suited up in clothes when they go to the beach. This is where sunscreen comes in.

Sunscreen is a double-edged sword. On the one hand, it deflects damage from UV rays. Yet there are downsides. One is that it gives you a false sense of security. Typically, no one applies enough sunscreen or reapplies it as often as needed. And even if you are a master sunscreen applier, it can't protect you from rays completely.

The other downside is that many sunscreens have chemicals—usually oxybenzone, avobenzone, octisalate, octocrylene, homosalate, and/or octinoxate—that, even though applied topically, are small enough to be absorbed by your system. One study found sunscreen chemicals in breast milk, amniotic fluid, urine, and blood plasma, a sign that they're making their way into the body. And by this time, I think you know what that means: your glutathione machinery is going to have to work extra hard to get those compounds out of your system, leaving you with less GSH to fight your body's other battles. There is also an environmental consideration here. Some sunscreen chemicals are known to harm coral reefs. Hawaii has become the first state to ban sunscreens with oxybenzone and octinoxate, two ingredients believed to contribute to coral bleaching. For the safest choices see "Sun Smarts" below.

SUN SMARTS

Here are some ways to reduce your exposure to glutathione-sapping UV rays.

▷ Wear protective clothing, including a hat and sunglasses.
▷ Hang out in the shade.
▷ If you exercise outside, work out in the early morning before the sun comes up or in the evening when the rays are substantially diminished.
▷ Choose a safe sunscreen: ones made WITH the mineral zinc oxide (avoid titanium dioxide, which doesn't degrade) and WITHOUT the oxybenzone, avobenzone, octisalate, octocrylene, homosalate, and/or octinoxate.
▷ And if you do get a sunburn? Apply some topical glutathione twice a day, which will take away the pain in 20 minutes, and help your skin heal much more rapidly.

ADDRESS STRESS

Staying serene and curtailing your reactions to the disquiet that surrounds all of us on a daily basis can help conserve antioxidants, including glutathione. There are, of course, many ways to de-stress: taking a walk, lounging in a hot bath, doing yoga. Some people do crossword puzzles or read mystery novels. Others binge watch or go bowling. Whatever works for you.

But on a daily basis, I believe meditation is the best way to keep stress from causing your emotions (and the subsequent toil they take on your body) to boil over. Meditation can help you temper anger and calm anxiety, and frees your mind of clutter, allowing you to return to your day refreshed. And all it takes is about three to seven minutes, two times daily.

During this week, set aside two periods for an easy meditation. You can be anywhere while doing it (well, not while driving) and either be sitting on the floor or in a chair. Here's all that's required of you:

While seated, rest your hands on your knees with your palms up. Tilt your chin up slightly and close your eyes. Now, take twenty-one breaths, inhaling slowly, then exhaling slowly. Focus on the spot between your eyes, letting your thoughts drift by without paying attention to them. That's it! Just twenty-one breaths and you'll feel a world of difference.

Note: I personally like it to be completely quiet when I meditate, but some find that using a mantra during their meditation is quite helpful. Mantras—the repetition of a phrase of two or three words—can help you detach from distracting thoughts, allowing you to unplug from the busyness of your mind. You can practice different methods to find what works for you—some people only need to focus on some white noise like a ceiling fan or calming music in the background to help them avoid distracting thoughts. A mantra, if you choose one, can be something like "I am loved" or "All is well." Whatever your method for unplugging, you will find that it does wonders for your overall calmness, clarity, and peace of mind throughout the day.

DETOXING WITH GLUTATHIONE

Your body is detoxing twenty-four hours a day—and it's pretty good at it, handling everything from by-products of physiological processes, to chemicals you may have ingested or breathed in, to metabolized food and drink (particularly alcohol). But there are so many toxins in our environment now that the liver and its aide-de-camp, glutathione, may need some extra assistance to ensure that all wastes and chemicals safely exit the body.

The steps I outlined in Action Plan One are a good start: They will help improve your detoxing power by increasing GSH. This is the best way to detox—better than any three- or five- or seven-day cleanse, juice fast, or other special diet. Those types of plans may help you feel good, but they don't really assist the liver in doing its job. You need glutathione for that.

I often hear about people doing things like colon cleanses as a way to detox their body. That's not detox—that's plumbing cleanup. It's like when my wife takes all the junk out of its hiding places and puts it in the garage for me to get rid of. It's taking out the trash. Detoxing is more complicated because it involves not just shuttling things out the door, but making the cleansing machinery—the liver as well as the gallbladder and bile—work properly. Glutathione does that by both reducing oxidative stress and transforming harmful substances so they can be easily expelled by the body.

There are also some other ways to encourage the body to get rid of toxins, and I will include them in this Action Plan. The plan is really two plans: If you're feeling subpar, low-energy, and/or have simply been eating and drinking too much, follow the first protocol (it's great for a postholiday cleanup or even a preholiday setup!). If you believe (or tests have shown) that you are battling a high level of toxicity, try the second. Each protocol recommends safe options and, as with most things in life, some of them will work better for some individuals than they do for others. The key is to find what works best for *you*.

PROTOCOL 1: LIGHT DETOX

If you feel "off," not at the top of your game, have gained weight, look pasty, or feel sluggish and bloated because you've been filling your body with crap, this plan is for you. It consists of two goals. The first is to revitalize you; the second is to get you on a path that requires no special detoxing at all.

I am a prime example of someone who was an on-and-off-again detoxer. I did my first detox diet many years ago and I felt fantastic. I said to myself, "Oh my goodness! This is how the human body is supposed to feel!" I felt amazingly good for a month and a half. Then, I went back to my old habits. The next year, I detoxed again, and it inspired me to eat healthier foods for a few months. Along came January and I still felt fantastic—I was eating well all the way till summer. Then, back to my old ways once more. The third year, my detox led me to eat well until Thanksgiving, then between Thanksgiving and Christmastime I ate whatever I wanted (and what I wanted wasn't good). Finally, I decided I'd had enough. Why just be healthy some of the time when you can be healthy (and feel great because of it) all the time? I wanted to feel high-energy and robust all year round. So, I stopped detoxing and instead just ate very close to The Glutathione

Diet in Action Plan One twelve months a year. And I'm the healthiest I've ever been.

As you prepare to try the glutathione approach to detoxing, it's a good time to ask yourself whether you want to be healthy all the time or just some of the time. Because, although we don't know all the answers, it's not likely that a week or two of detoxing can make up for unhealthy habits the remainder of the year. Consider, for example, that it takes about forty days for cells to replenish themselves. If you're eating poorly, drinking a lot, not exercising, those harmful effects can last longer than you might expect.

So, going forward, think of detoxing as both an end to destructive habits and a beginning of full-time, year-round healthy living.

Step 1: Follow Action Plan One

Everything in this plan will increase your glutathione levels and that's the best way to detox.

Step 2: Add Liver-Nourishing Foods and Food Extract Supplements to Your Diet

In addition to the list of GSH-raising foods I shared in Action Plan One, these foods have known properties that promote good liver health. Except coffee (limit to a cup a day), sprinkle them liberally into your diet.

Artichokes—Properties in this vegetable have been shown to help prevent toxins from entering the liver cells.

Coffee—Coffee drinkers are less likely to have liver cancer and non-alcohol-related fatty liver disease.

Dandelion greens—These astringent greens have long been used to promote liver health. Toss them into salads.

Brussels sprouts—A substance in cruciferous vegetables (many others are included in The Glutathione Diet) contain quercetin, which helps fight liver dysfunction.

Turmeric—This spice is an anti-inflammatory, rich in antioxidants, and known to protect the liver. You may see it referred to as curcumin, which is the active ingredient sometimes extracted and sold on its own. Add turmeric to your meals once a day (great in stir-fries and egg dishes; many cafés also now sell turmeric lattes). If this isn't possible, consider a turmeric supplement. Look for one that includes such digestive aids as bioperine or piperine (black pepper extract) that help enhance absorption.

Milk thistle (silymarin)—Supplements made from this herb have been shown to help heal liver damage from exposure to chemicals. If you decided to try them and are taking any other medications, check with your doctor first to make sure the milk thistle won't interfere with the other drugs.

Cordyceps—This mushroom has long been used in Chinese medicine to promote immune health. As a detoxer, I've found some animal research showing that Cordyceps can increase glutathione. You can find Cordyceps supplements (pills or tinctures; some are powders you can mix with water) on store shelves, or try Cordyceps hot tea bags.

Step 3: Supplement with Glutathione

As noted in Chapter 8, there are many ways to supplement with glutathione, from oral supplements to under-the-tongue drops to skin creams. I recommend that you follow the instructions on the label, then see what kind of results you get. While you may not be able to see if your body is expelling more toxins, you will know how you feel and whether any health problems you are struggling with clear up. That's a good way to see whether the GSH is having an effect.

If you are interested in trying our Auro GSH gold solution, the protocol for detox that we used on thirty-four patients in the clinical trial I wrote about on page 85 was 0.5 ml (that's four squirts), twice a day. I recommend this same regimen with continued use until symptoms abate. Keep in mind that eliminating heavy metals, in particular, is a very slow process and can even take one to two years.

Another option is IV glutathione, but that is time-consuming and expensive since you will still need to supplement for about a year before you see a change in your heavy metal blood test results. Using topical glutathione daily will give you better results, even if it does take longer. If you do decide to try IV glutathione, choose a reputable provider, preferably one that is run by medical professionals.

Step 4: Unplug

Some of the things that we love dearly (and dare I say are addicted to), such as our smartphones, computers, tablets, and everything else with a personal screen, can contaminate our body in ways that we don't notice. These devices create electromagnetic fields (EMFs), a type of energy that has a chemical-like effect that experts are predicting will cause cumulative damage in generations to come. Electromagnetic fields are also generated by electrical appliances and electric power lines, but the devices we hold so dear (and hold so close to our body) have made the danger more immediate. We already know that even in the short term, EMFs cause oxidative stress, and that the overproduction of free radicals puts a strain on the body's glutathione defenses.

I know that nobody is going to give up their cell phone, but if you are trying to detox, I recommend that you try to minimize your EMF exposure. There are little ways you can do that. Might I suggest, for instance, that you do not sleep with your phone or any other technological devices on your bedside table. As a matter of fact, I advise that you don't charge your cell phone or computer in your bedroom at all. Make it a device-free area. (Just for the record, I'm also not a fan of having a television in the bedroom, because it's too tempting to stare at the screen for hours on end, which has proven to disrupt sleep patterns and does not promote the good night's rest that's essential for good health.)

I also suggest scheduling in "technology detoxes"—times when you unplug for at least an hour a day, or keep chunks of your weekend device-free. This not only limits EMF exposure, it also promotes live human interaction and connection, which is just as important for your well-being as anything. In my family, we keep a basket in the middle of the dining table where every family member must place their devices at shared mealtimes. We also have many other times in the Patel family home that are designated as "device-free." Try this; you'll thank me later.

HOW TO MINIMIZE DETOX SIDE EFFECTS

EVEN IF YOU ARE only doing a light detox, eliminating certain foods and adding in others (even if they are way more nutritious) can still be a shock to the system. You may be surprised to find that you feel achy, irritable, anxious, headachy, and/or experience nausea. Other side effects can include brain fog, constipation, itchy skin, rashes, fatigue, and sleep difficulties. *Hey,* you'll find yourself thinking, *I thought this was supposed to make me feel better!* Don't worry; in time it will. But at the beginning, your immune system may react in ways that intensify inflammation for a few days.

If you find that the effects of a detox are extreme—some people, for instance, experience a detox rash caused by toxins exiting through the skin—you might back off your protocol for a bit, then once the rash dissipates, return to the plan. To lessen a rash, keep your skin clean and, if needed, moisturize with coconut oil.

Being prepared for some side effects can help make your detox successful. So can planning ahead so that you have all the foods and other detox items you need on hand before you start. It's also always best to start a regimen on a nonbusy weekend, and at a time when nothing major or stressful is going on at home or work.

A few more tips for easing any detox discomfort:

> ▹ Take 400 mg a day of magnesium citrate powder or supplement. This helps with elimination, sleep, aches, pains, and stress. Check with your physician first, but you can take a higher dose if you're struggling with constipation.
> ▹ Drink lots of water—at least 8 to 10 glasses a day. Water helps flush the body clean.
> ▹ Make sure you get a full night's sleep. Your body goes into full-repair mode while you're sleeping so get plenty of shut-eye.

▷ Exercise every day, even if it's just a walk around the block or some yoga stretches before bed. Keeping your circulation moving helps to eliminate toxins from the body.

▷ Soothe yourself with hot herbal tea and warm Epsom salt baths.

▷ Most important, don't be hard on yourself. You don't have to be perfect. Everything you're doing is contributing to your health and you're to be congratulated for that. Being kind to yourself will make the process go smoother.

⋯⋯⋯●⋯⋯⋯

PROTOCOL 2: HEAVY DETOX

Knowing when you need a heavy detox is a little tricky since you can't always test for it. True, you can be tested for heavy metal contamination through a simple blood test, and I recommend this test if you think you've been exposed to heavy metals or your symptoms are otherwise unexplainable. However, your body may also have been exposed to other types of toxins that don't show up as well. Remember Bryanna (page 69)? She had no idea that she was poisoning herself with overuse of hair dye. Some of the symptoms that may indicate that you need a heavy-duty detox include:

* Extreme fatigue
* Sudden allergies
* Headaches
* Digestive issues
* Muscle aches and pains
* Brain fog

This detox protocol includes the lighter protocol, then takes it a few steps further. One thing to remember is that most times your symptoms won't disappear overnight. You may need months to get your body back to its most vibrant self.

Step 1: Follow the Light Detox Protocol

Step 2: Add Supportive Supplements

Glutathione has superpowers—but even Superman needed help from Lois Lane. When you're doing a heavy detox, I suggest adding in nutrients that work in tandem with GSH. For this type of detox, supplements are your best bet for getting in those nutrients, but if you're supplement-averse, I've added some of the top food sources. The mighty five:

1. Vitamin E. If you're following Action Plan One, you will have noted my recommended supplements (page 154). For this purpose, I suggest the same protocol: 400 mg of mixed tocopherols daily, preferably Unique E brand, which has perfected the mixed tocopherols. Food sources: wheat germ oil, sunflower seeds, almonds, hazelnuts, peanuts (and peanut butter), spinach.

2. Chromium picolinate. This supplement also works hand in hand with glutathione to keep all detox systems go. It helps nourish the skin too. (See page 178 for more on this important mineral.) I recommend two 200 mcg capsules of chromium picolinate, two times a day, then discontinuing after nine to twelve months (your body will have built up enough of it by then). Food sources: broccoli, potatoes, whole wheat bread, garlic, basil.

3. N-acetyl cysteine (NAC). Earlier in this book, I talked about how the body builds glutathione using three amino acids: cysteine, glycine, and glutamate. We need all three. It's like going into your kitchen to make bread from flour, water, and yeast (and a little salt). If these ingredients aren't added in the proper proportions, mixed together well, then baked in the oven, you will never have bread.

In the case of glutathione, one of those necessary ingredients—cysteine— can be rather elusive, which is why N-acetyl cysteine, a supplement form of cysteine, is a good option when you're in the heat of a detox. On its own, NAC has been linked to positive outcomes with chronic respiratory conditions, fertility, and brain functions. There are also numerous studies on NAC that show it can positively affect depression, anxiety, OCD, and addiction.

Most studies use a dose of between 1,200 to 2,400 mg of NAC, taken in 600 mg doses throughout the day. One study found that this dosage increased glutathione levels by 30 percent in a seven-day span. Food sources: NAC is not found in foods, but cysteine is. See "Foods with the Highest Concentrations of Cysteine," page 143.

4. Copper. One of the good metals we need in our body (see page 178 for an explanation on good metals), copper is only required in trace amounts—getting too much of it can cause health problems, so only take up to 30 mg of copper a day (the RDA is 200 mg, but you'll get some through food and, possibly, your local tap water). Food sources: sesame seeds, unsweetened chocolate, potatoes, cashews, sunflower seeds, mushrooms.

5. Zinc. This mineral helps remove heavy metals from the body. Add in 2 mg of zinc a day. Take your copper and zinc at the same time, in the morning. Food sources: baked beans, pumpkin seeds, cashews, chickpeas, oatmeal.

Optional:

Alpha lipoic acid (ALA), a powerful antioxidant in its own right that helps your body produce glutathione. Most studies on ALA use a 600 mg per day dose. ALA is made by the body; the small amounts of the antioxidant in food aren't well absorbed by the body, but supplementary ALA is.

B vitamins. A good B complex vitamin, taken daily, will help you maintain your body's vitamin and mineral stores, and support your liver's function. You can choose either cobalamin or methylcobalamin B complex, but I prefer the latter. Food sources: B_{12}—nutritional yeast, trout, salmon, yogurt, egg; B_6—chickpeas, salmon, potatoes, bananas; folic acid—spinach, black-eyed peas, asparagus, Brussels sprouts, avocado.

Step 3: Try Additional Methods of Kick-Starting the Body's Clean-Out System

The following are a few methods that may increase your glutathione production and help speed detoxing.

Cryotherapy. Cryotherapy, or cold therapy, was once reserved for doctors' offices. Now, though, it's available at corner wellness centers in many cities. Cryotherapy involves applying subzero cold to the body, either the whole body or particular areas, for short periods of time, using nitrogen gas to cool the air. Many people seek it out to help with workout recovery (it's very popular with athletes), but it's also used for everything from arthritis and mood enhancement to slowing aging of the skin. One way it works is by stimulating the central nervous system, releasing endorphins to help with pain management. It's also thought to work by causing the body to draw blood to the core (it does it to protect your vital organs from the cold), where it also gets the opportunity to stock up on oxygen, enzymes, and nutrients that are vital to cell regeneration.

How does it relate to detoxing? Some research shows that cryotherapy can increase levels of antioxidants in the body, including glutathione. One study by Polish researchers found that twenty consecutive days of three-minute cryotherapy sessions increased glutathione in thirty young men. One possibility of the bump in GSH is that the cold temperature shocks the cells into action, spurring them to increase the enzymes that process glutathione, so that there's more of it.

Infrared saunas. If cold is good, what about heat? Turns out it, too, may have detoxing benefits. Infrared saunas, popular now at spas, use infrared light to heat the body from within, rather than heating the air around the body, as a traditional sauna does. During infrared sauna sessions, cells release toxins, which are then swept out through pores as you sweat. This is not the same as liver-glutathione detoxification, but it is a good adjunct to clearing the body of wastes and other substances it needs to get rid of.

An infrared sauna session also raises your core temperature, inducing an artificial fever. A fever is the body's natural mechanism for strengthening and accelerating the immune response. This enhanced immune response, combined with improved elimination of toxins and wastes via sweating, increases your overall health and resistance to disease. I suggest that if you do go in for an infrared sauna session that you spray a topical form of glutathione on after the session to help expedite the detoxification process even further.

Sulfur bathing. Something that I recently put on my family's bucket list is a trip to Iceland to visit the sulfur-rich geothermal pools. The specific one that I've got my heart set on is the Blue Lagoon in southwestern Iceland. It's been nicknamed the fountain of youth for its waters and is labeled one of the twenty-five wonders of the world. The lagoon sits at 39°C (102°F) year round, making it the perfect temperature for bathing.

Bathing in sulfur mineral waters is a time-honored health tradition; people have been doing it for thousands of years. And for good reason. Sulfur can be absorbed through the skin and have an effect on what's happening internally in your body. This potent mineral, as you may remember, is a component of cysteine, one of the building blocks of glutathione. And according to a paper published by Spanish researchers in 2017, sulfur bathing is "starting to show benefits for pulmonary hypertension, arterial hypertension, atherosclerosis, ischemia-reperfusion injury, heart failure, peptic ulcer, and acute and chronic inflammatory diseases."

Fortunately, you don't have to go all the way to Iceland for a sulfur bath. There are mineral baths all over the country and some local spas offer sulfur-based hydrotherapy.

Intermittent fasting. Intermittent fasting (IF), as its name suggests, involves switching between periods of eating and fasting. When you eat, you eat a normal amount of food, but just within a smaller time frame, which is called your "eating window." It's not for everyone, but is simply another option for reaching health and, especially, weight goals.

The whole idea behind intermittent fasting (and why I think it's compatible with detoxing as long as the foods you choose adhere to The Glutathione Diet protocol) is to give your digestive system a break, and allow your body to focus on all of the other tasks and cellular functions at hand. When you fast on a recurring basis, you don't spend energy digesting food. It also cuts down on any waste products created by the digestive system that require glutathione assistance to be removed from the body.

I am not an expert on IF, which is why I encourage you to look elsewhere for specifics on how best to do it, but I did want to make you aware of the eating regimen and tell you about some of the ways its effects coincide with glutathione's tasks in the body, including helping to reduce insulin resistance, reducing "bad" LDL cholesterol and blood triglycerides, and stimulating weight loss.

WHAT IS CHELATION THERAPY?

WHEN A PATIENT IS shown to have metals like mercury, lead, or arsenic in their system—as determined by a doctor-ordered blood test—health-care providers often prescribe what's known as chelation therapy. To chelate is to form a bond with a metal atom. Chelation therapy uses special drugs that bind to metals in your blood so that they can be excreted from the body. You get the chelating medicine through an intravenous (IV) tube in your arm or through a pill. Once the drug has attached to the metal, your body removes them both through your urine.

This type of treatment is by prescription only and should be closely monitored by a physician. There are certain risks involved, including the depletion of important nutrients, such as calcium, copper, and zinc. Kidney damage is a risk and so are side effects like headache and vomiting. I am not advocating chelation for everyone, but in serious cases, it might be recommended by your doctor. And when it is recommended, I suggest using it in combination with either an absorbable form of glutathione or daily GSH IV treatments. This will help rush the poisonous toxins that have been stirred up by the chelation process out of the body.

OPIOID DETOX—CAN GLUTATHIONE HELP?

As WE ALL KNOW, opioid addiction is an epidemic in this country. It's a national crisis and painful to watch, especially when it involves an overdose. Emergency medical people generally treat opioid overdose with a drug called naloxone. An overdose will cause respiratory failure—people simply stop breathing—and naloxone can reverse that immediately. As of right now, naloxone is approved only to reverse opioid OD, but I believe that topical and IV glutathione should also be carried in every ambulance and stocked in emergency rooms and detox centers around the nation. And I'll tell you why: Many opioid deaths are due to acetaminophen toxicity.

Acetaminophen, which is sold over the counter as Tylenol, is also frequently combined with opioids and marketed as a pain reliever (Vicodin, for instance). Although the FDA has limited how much acetaminophen can be combined with prescription drugs, it can still pose a risk when someone takes too much of the drug.

Acetaminophen may seem innocuous—you can get it at any store, after all—but it can be very toxic to the body in large doses (see page 23 for a fuller explanation). It also depletes glutathione so completely that there's little left to push the opioids out of the body. But once GSH is infused back into the system, detoxification resumes, potentially stopping an acetaminophen overdose in its tracks. For that reason, I hope to one day see high-dose IV glutathione and topical, absorbable forms become part of the treatment for opioid overdose.

—GSH FOR BETTER,
BRIGHTER SKIN

THE BODY IS AN INTRICATE SYSTEM OF LINKED BIOLOGICAL HIGHWAYS
and byways—everything is connected. And that is especially true when
it comes to the skin, which reflects what's going on in other areas of your
body. So, if you follow Action Plan Two for detoxing because you are, say,
fatigued and experiencing digestive issues, your skin is going to benefit too.
Likewise, The Glutathione Diet in Action Plan One is going to make your
skin more luminous. When your body is clean on the inside, it reflects radi-
antly on the outside.

This is my way of saying that each plan in this book, by boosting glutathione, will have a payoff for your skin. Of course, there are also specific things
you can do to enhance the "mechanics" of your skin—the cell turnover,
collagen creation, and pigmentation production, for instance—in order to
slow down aging and brighten your complexion.

I believe in keeping skincare regimens relatively simple, basing them on
the following four cornerstones:

1. **Eat well.** What you feed your stomach also feeds your skin. Wholesome, glutathione-building foods (for a refresher, see page 142) will nourish your skin while junky, unwholesome food will sap it of vibrancy.
2. **Keep your body clean and detoxed.** This includes exercising and sweating, which we discussed in Action Plan Two (page 159). Cryotherapy and infrared saunas are also recommended.
3. **Exfoliate.** The skin has a nifty way of replenishing itself by sloughing off dead cells. But as you age, the process slows down. Give it some help by exfoliating (see page 175 for more info).
4. **Treat with glutathione.** GSH is a major player in both protecting the skin and repairing skin damage. It's essential for stamping out free radicals, building collagen, and lightening dark spots. Simply raising your glutathione levels by natural means (see Action Plan One) will help, but for more significant results, you'll need to use topical glutathione.

STEPS FOR SKIN RENEWAL

These four cornerstones of skincare call for a look back at Action Plans One and Two. To eat well, I suggest The Glutathione Diet in Action Plan One. Actually, following every recommendation in the first plan can help with skin renewal, but at the very least consider following the dietary and sun protection recommendations. Action Plan One will also help with detoxing, lowering the number of contaminants that get into your system, and helping you expel the ones that do. For more intensive measures, check out Action Plan Two.

The next two steps involve exfoliating and treating the skin with glutathione. Here's the lineup.

Step 1: Follow The Glutathione Diet and Sun Protection Recommendations from Action Plan One

This is skincare both from the inside out and the outside in.

Step 2: Follow the Light Detox Steps in Action Plan Two (Optional)

If you are dealing with symptoms other than lackluster, aging skin—such as fatigue, gastrointestinal issues, general malaise (that is, you just don't feel right)—a detox will be especially helpful. Your skin will benefit and so will the rest of your body.

Step 3: Exfoliate

Every thirty to forty days, your skin gets rid of old surface cells and pushes new ones to the top. As you'd expect, the new ones are plumper and shinier than the old ones, so you want that turnover to happen as regularly as possible. But the process slows down as you age; that's one reason that older skin lacks that certain glow that young skin has. (See Chapter 5 for more on the changes that come with aging.)

One way to help skin turnover along is by exfoliating: removing the dead cells to make room for the new. When you look in the mirror, many of the cells on the surface of your skin have died and served their purpose. And yet they hang on for dear life, clogging pores and making your complexion look dull. They're going to slough off eventually, but if you can give them a shove, your skin is going to look brighter for it. And by getting rid of this cellular debris, you also clear the way for topical treatments, including glutathione, to penetrate the skin.

There are two ways to exfoliate. One is with products that contain chemical ingredients that dissolve the dead cells. The other is through physical exfoliation. This can be accomplished by using a product with small grains in it that break up the cells so they can be washed off, or by using something, such as a rough cloth or sponge, to accomplish the same thing. Physical exfoliation can be hard on the skin, so I urge you to be gentle if you take this approach.

My preference is for the chemical approach. There are several different active ingredients that do the job. Here are the most effective ones:

Glycolic acid. The number one exfoliant that I recommend is the alpha hydroxy acid (AHA) called glycolic acid. Don't let the word *acid* scare you—glycolic acid, derived from sugarcane, is safe for the skin. Since it's a small molecule, it penetrates deep into the top layer of the skin, making it an effective exfoliator.

Lactic acid. If your skin tends to be sensitive, try this AHA, a milder exfoliator than glycolic acid but still effective. This is also a natural acid derived from plants or milk.

Salicylic acid. Classified as a beta hydroxy acid (BHA), this acid is often used to treat acne.

Black charcoal. In addition to exfoliating properties, black charcoal has been shown to draw out impurities from the skin.

Honorable mentions. A few other ingredients that can help with exfoliation are fruit enzymes (e.g., papaya and pineapple extracts) and vitamin C. Along with aiding in exfoliation, vitamin C also has antioxidant benefits.

In general, choose products that suit your skin type and are made without parabens, phthalates, or petrochemicals. How often should you exfoliate? Begin with once a day, which is optimal. If you notice irritation (redness, inflammation), cut back to two to three times per week. The best time of day to exfoliate is in the evening before bed, because (a) you can remove the debris that's accumulated during the day and (b) the body has a surge of cell turnover while you're sleeping.

Also consider taking exfoliation beyond your face and neck and exfoliating the rest of your body three times a week. There are body-exfoliating products on the market similar to the ones available for your face as well as physical exfoliators, such as loofahs, Japanese scrubbing cloths, and a technique called dry brushing that's very effective. Dry brushing is exactly what it sounds like: Stroking the skin with a very soft brush helps slough off dead cells and liberate the plumper new cells underneath.

THE SAFE WAY TO EVEN, SPOT-FREE SKIN

PIGMENT DEPOSITS ARE GENERALLY thought to be caused by sun exposure ("age spots" are really cumulative-damage sun spots), genetic predisposition, hormone changes, and skin irritation. Men can have them, but they're most common in women, sometimes triggered by pregnancy or hormonal changes and imbalances, a condition called melasma. The spots normally appear on the face but can also appear anywhere on the body.

Lightening hyperpigmentation spots was one of the topics of the discussion I had with Howard Maibach, MD, who mentored me a few years back after I was introduced to him at an event. Dr. Maibach is a well-respected professor of dermatology at the University of California, San Francisco, School of Medicine.

Dr. Maibach was very intrigued by what he'd heard about glutathione's skin tone–evening capabilities. He was already very familiar with the many roles glutathione plays in the body, but the skin-evening aspect was new to him. It was the jumping-off point for our many discussions at his beautiful San Francisco home, where we pondered the abilities of GSH. Dr. Maibach was involved in the original critical thinking process and approval of a product called hydroquinone back in the mid-1960s. It was eventually taken off the market in some countries when fears that it increases the risk of cancer became too great. There hasn't been anything else really effective on the market since. As a result, Dr. Maibach has been in constant pursuit of a better and safer product for hyperpigmentation. Now he was learning about glutathione.

For some time, glutathione has been well known among celebrities and in Asia for its skin tone–evening capabilities. Some of my Chinese colleagues have performed studies on the antioxidant's effects on melasma and other skin discoloration, using the solution developed by my company. I must say that the before and three-weeks-after pictures generated by the Chinese research were quite shocking— and exciting. It spurred us to go ahead and patent our topical glutathione in China.

The only other well-known spot lightener on the market, besides hydroquinone, is a natural ingredient called kojic acid, which comes in different forms, and has varying results. Kojic acid is actually a chelation (metal-binding) agent.

It's produced by several different species of fungi and is a by-product of the fermentation process of a type of rice used for the manufacturing of sake, the Japanese rice wine. I, you will not be surprised to learn, am partial to glutathione for erasing sun spots and melasma, but it's good to know what's out there so you can make an informed decision.

........●........

A SKIN-ENHANCING SUPPLEMENT

CHROMIUM IS A MINERAL that helps the body process fats, carbohydrates, and protein. Chromium, as it turns out, also helps the body combat melasma and hyperpigmentation. Research also shows that chromium picolinate (a supplemental form) can reduce insulin resistance and help lower the risk of cardiovascular disease and type 2 diabetes. So, while you're beautifying your skin, you'll also be protecting your health.

The body needs certain essential metals for overall good health. I know that sounds surprising given that we usually want to get metals *out* of our system, not in, but we actually do need very small quantities of chromium along with manganese, vanadium, iron, cobalt, copper, zinc, selenium, strontium, and molybdenum. Interestingly, a deficiency of these essential metals may actually increase susceptibility to heavy metal poisoning.

I recommend this supplement in Action Plan Two, but in case you're not following it, let me repeat the suggested amount: 200 mcg capsules of chromium picolinate, two times a day, then discontinuing after nine to twelve months. The chromium will build up slowly, allowing you to get to a healthy level without getting too much of this essential mineral. If possible, ask your doctor to prescribe prescription-grade chromium picolinate. You can also find high-grade versions online or at such stores as Whole Foods.

........●........

Step 4: Treat with Glutathione

In Chapter 7, I outlined all the ways that glutathione fights signs of aging in the skin, including significantly lightening sun spots. The glutathione we make on our own is a skin savior, but—and this is a big *but*—we need a lot of it, often more than we can make, to address such skin conditions as acne or to repair damage from the sun. That's where topical glutathione comes in.

Many topical products are only skin-deep—that is, they don't penetrate the different layers of the skin to get into the cells that need their help. Since our skin is the largest organ of the body as well as the gatekeeper for much of what gets into the body, this is actually a good thing. You don't want everything you come into contact with to be absorbed by your skin. But when you are trying to introduce something beneficial, this impenetrability can be frustrating.

As I've mentioned, one of my professional missions in life was to find a way to make glutathione absorbable, opening up the possibility that it could be used to treat every condition that had roots in a lack of GSH. So, before I give you a prescription for using topical glutathione to rejuvenate your skin, let me explain how it gets in there to do its job.

WORKING BELOW THE SURFACE

There are two main ways that active ingredients in a cream, lotion, or spray can penetrate the surface of the skin and reach the lower layers. There are openings (pores) all over the body, but the channels leading to them are not the same. One is for lipids, especially sebum, which is an oily substance that moisturizes and protects the skin. The other channel is for water, or perspiration, which is released to cool the body. Either channel can be used to deliver ingredients deep into the skin. However, they're not equal. The water channels of the skin are tiny, and few and far between, but they function as a highway. The lipid channels, while bigger and more common, are sticky and slow. Thus, the water channels are a far superior road to take when you want to deliver something to the farther reaches of the skin where it can actively assist in reparative processes.

The topical, spray glutathione we use in our products is water soluble, which is why it works so well. We also have glutathione-based skincare lotions with the same water-soluble form of GSH, plus a patented complex with the antiaging, collagen-building, moisturizing, and antioxidant ingredients vitamin C, hyaluronic acid, CoQ10, cycloastragenol, carnosine, dimethylaminoethanol (DMAE), and mixed tocopherols. These are all ingredients I recommend you look for in any skincare products you use.

WHAT INGREDIENTS TO LOOK FOR IN SKINCARE PRODUCTS

AS INSTRUMENTAL IN BEAUTIFYING skin as glutathione is, no one ingredient can provide everything that the skin needs. The trouble is, *so many* ingredients are used in skincare products that it's difficult to know what you need. When we were formulating our products, we did extensive research on what ingredients best enhance and de-age the skin. Here are the seven I believe are most effective and what they can do for you.

Vitamin C—As a fellow antioxidant, vitamin C shares some of the properties of GSH: It fights the free radicals that age skin, has anti-inflammatory properties, and helps reduce hyperpigmentation. Vitamin C is also an essential player in collagen and elastin production, one reason it's been shown to reduce wrinkles. If applied before sun exposure, it protects the skin from photo damage as well.

Hyaluronic acid—One of the many natural properties in the skin that decreases with age is hyaluronic acid. This molecule is a humectant; that is, it traps water in the skin, giving the complexion a plumper and dewier appearance.

CoQ10 (also called coenzyme Q10 and ubiquinone)—We naturally produce CoQ10, but production slows as we get older. This critical enzyme enhances the work of antioxidants, keeps the mitochondria running properly, which helps the skin repair itself, and retards the breakdown of collagen.

Mixed tocopherols (compounds containing Vitamin E)— The antioxidant vitamin E not only helps stamp out damaging free radicals, it has emollient properties that hydrate and smooth out rough skin.

Honorable Mention

These may be a little harder to find in skincare products as they're not widely used at this time. But the research behind them is good and I believe they are important ingredients.

Cycloastragenol—This antioxidant has been shown to reduce signs of aging, possibly because it encourages telomere growth, prolonging cell life.

Carnosine—A substance that inhibits the formation of advanced glycation end products (AGEs). AGEs are known to reduce the elasticity (bounce-back) of the skin.

DMAE—Known to firm muscles lying beneath the skin, somewhat in the vein of Botox. The effect is to make the skin look smoother.

............●............

GLUTATHIONE PROTOCOL

Topical glutathione can be used to spot-treat skin problems, such as sun spots, or to rejuvenate the whole face, neck, and décolletage. My colleagues and I often joke about how you can apply glutathione just about anywhere for anything. It's like that funny running joke in the film *My Big Fat Greek Wedding*. Throughout the movie, the father of the bride spritzes everyone and everything with Windex. "That'll fix it," he says. Same with glutathione.

Now that said, when it comes to caring for your complexion, I recommend the whole-face approach. If you are using a topical glutathione lotion or serum, follow the directions on the package. If you opt for my Auro glutathione spray for topical use, here's how to do it:

Spray the glutathione in the palm of your hand, then use your index finger to spread it all over your face, neck, and, if you like, décolletage. Exclude the under-eye area and lids, but dab the glutathione onto the lines in the corners of your eyes. You can also use the same technique to apply the solution to the back of your hands. After thirty minutes, wash off the glutathione with warm water, and then apply your other moisturizers and serums. Use the glutathione once a day or every other day, morning or before bed, whichever fits in best with your schedule. (Note: If you find the sulfur smell of the solution too unpleasant on your face, spray it on your abdomen—since it works systemically, it will affect your skin no matter where you apply it.) If you're using it in the morning, add sunscreen to your regimen either before moisturizing or, easiest, using a combination moisturizer/sunscreen.

If you want to take a spot treatment approach, follow the same daily regimen, but simply apply the glutathione only on the places (such as sun spots or pimples) that need it. This is also a good rule of thumb for sunburns, bruises, and scars. For a sunburn, in particular, apply glutathione every three hours to reduce the pain.

OTHER SKINCARE USES FOR TOPICAL GLUTATHIONE

GLUTATHIONE HAS BECOME A popular tool for skincare professionals. Mimi T., an aesthetician in Newport Beach, California, with a wide array of celebrity clientele, has been using topical glutathione solution as an adjunct to microneedling (a procedure using fine needles to puncture the skin and stimulate collagen building). The GSH helps further the effects of the procedure and speeds the healing process. The results she's seen with glutathione have led Mimi to begin developing a GSH-based product designed to plump up the skin and build collagen.

Some cosmetic surgeons are also using topical glutathione to promote recovery from both surgical and laser procedures. And sometimes patients are using it on their own when a procedure doesn't go right. For instance:

The use of lasers to even out skin color, diminish lines and wrinkles, and really do just about everything to improve the look of the skin is so common now that you can probably find the service only a mile or so from where you live. Most times things go well, but sometimes they don't. That's what happened to Jeong, a patient of mine. Here's what she told me: "I had no idea that after signing up for a certain anti-aging laser treatment with my doctor that was supposed to smooth and beautify my skin, that it would actually destroy my skin!"

I'm not sure exactly what went wrong, but you can imagine how terrible she felt when her face became rough, dotted with pigmentation spots, and a shade darker than it was before. Her skin even began to peel. She was upset but also too afraid to go through another procedure to try to remedy the situation. When Jeong read about topical glutathione online, she decided it might be a much better option. It restored her skin's texture and lightened the pigmentation. The moral here, I think, is that you never know what glutathione is going to do for you. There is little to no risk in using it to try to solve a beauty-related problem. In this situation, it literally helped Jeong save face.

I personally use topical GSH as an aftershave to help heal any nicks or sensitive patches on my face, and for overall antiaging and skin health needs. GSH can also be used to treat sunburns, heal cuts, and lessen the appearance of stretch marks. It can be effective in eradicating some cases of eczema, psoriasis, and rashes (depending on their source) too. In one case, glutathione was used to help a thirteen-year-old erase a case of eczema.

Ryland prided himself on being the healthiest among the three children in his family, so he was quite appalled when he woke up one summer morning with a rash that ran from his fingers all the way up to his elbows—on both arms. It was itchy and stinging and horrible. His mom took him to urgent care and Ryland was diagnosed with eczema, then prescribed a steroid cream. He then visited his regular doctor to get tested for just about everything that they could think of to find out what was causing the skin disorder. It turned out he had a few allergies, but even after removing the allergy-generating foods from his diet, the eczema would not go away. The steroid cream stopped working too.

So, one night his mom coated both arms with glutathione, and by the next morning, the rash showed signs of improving. With ongoing treatment, a week later it was gone. "I have no idea what glutathione is or how it even works, and I hate the rotten egg smell," says the thirteen-year-old. "But it healed me and made me really happy!"

········●········

AFTERWORD

A GLUTATHIONE-RICH FUTURE

With your interest in glutathione, you are on the forefront of revolutionary new pathways in health care. For the first time, it's possible to employ strategies to increase your personal GSH levels both through natural means and through the use of supplementary forms of the antioxidant that are bioavailable. Glutathione supplementation may have been around for a long time, but it's only now that we're finding ways to ensure that it really works.

That's one way we're blazing a trail to better health.

The other—and I believe even more exciting—way we're making groundbreaking advances is in the public health sphere.

For many years, my pharmaceutical company supplied IV glutathione to practitioners around the country. We were, in fact, one of the biggest suppliers in the US. But in November 2019, I decided to put that business on the back burner to free up time to begin working with various international health and philanthropic groups to get glutathione out to underserved populations throughout the world. I'll also be continuing to explore the power of glutathione and all the ways it can fight disease and help people achieve optimal health.

AREAS OF ONGOING RESEARCH

So many health fads come and go; sometimes they're even found to be quite bad for us. Do you remember the diet era when everything was all about avoiding fat? When more preservatives, chemicals, and sugars were added to replace the fat, that way of eating became something of a fiasco. Still, there are other fads that, because they're based on solid evidence, transition from being a novelty into being a necessity. That's what I believe is the case with glutathione. It may seem trendy at first, but going forward, research is going to show that it must and will be here to stay.

Of course, we still have a lot to learn about glutathione. I am most interested in looking into how it works in areas of the body that we haven't yet explored well and what glutathione can do in combination with other remedies. Here are some of the areas that I will be investigating in the future.

GLUTATHIONE AND VIRUSES

As I write this, in April 2020, we are smack dab in the center of a scary COVID-19 (coronavirus) outbreak. It is too soon to know what will happen—how long we will have to distance ourselves from one another, when business will return to normal, whether we will see a cure and/or vaccination for this.

Controlling both common influenzas and uncommon viruses is a new area of study for me and my research colleagues. Glutathione has already shown promise in treating the flu (see page 56), and I am hopeful that we will learn more about how to use it to fight other lethal viral strains, such as the coronavirus.

GLUTATHIONE AND HORMONE REGULATION

Hormones are active throughout the entire body, working as chemical messengers for countless tasks—they're critical to just about every aspect of your body's operation. I firmly believe that having your entire system in balance goes a long way toward keeping your hormones synchronized. But as we encounter environmental toxins and stressors, and especially as we age, a lot of us simply need some extra assistance in balancing our hormone levels.

Hormones exist in a very delicate ecosystem; the system is a beautiful symphony, but fragile. If one "instrument" is out of tune, the entire orchestra can be off-key. One of the worst side effects of out-of-balance or low hormone levels is emotional, mood, and energy complications. No one feels good when their hormones are out of whack. Women, for instance, can experience excessive premenstrual syndrome or a particularly tough ride through menopause. Men can end up with a lack of libido or impotence. And all genders can take a hit in the sleep, energy, cognitive performance, and immune health departments. Even glutathione is affected since some forms are regulated by thyroid and sex hormones.

Glutathione may also return the favor by helping to metabolize hormones in the liver, the major site for hormone metabolism—it's where they are broken down and then excreted out of the body. To me, this begs the question, Can glutathione correct hormone imbalances? This is a topic I'm excited about digging into with some of my research affiliates. Stay posted on this one.

NAD: A CRITICAL COENZYME

NAD, which stands for nicotinamide adenine dinucleotide, is a critical coenzyme found in every cell of your body. It's involved in hundreds of metabolic processes, including creating energy in the mitochondria. But like glutathione, NAD declines with age. How can we replenish it?

Even though NAD was originally discovered in 1906, interest in the coenzyme has jumped considerably. Within the last few years, several NAD supplements utilizing an ingredient called nicotinamide riboside have popped up on the market. It's being used for everything from increasing athletic stamina and treating chronic fatigue to helping fight Alzheimer's and lowering cholesterol. Like glutathione, it's multipurpose. But I have reservations about what's out there now. My research partner, Dr. Chinh Tran, and I are looking to create something even more active and bioavailable than these current supplements. Ultimately, we hope to find a good companion to glutathione, and so far, the findings in our lab on NAD have been very exciting.

GLUTATHIONE AND CBD

Unless you've been hiding under a rock lately, you've probably heard of CBD, an abbreviation for cannabidiol, an active compound in cannabis and all types of non-THC hemp plants. CBD is in almost every kind of product you can think of now, from gummy candies and coffee to creams and bath bombs. Of course, you can just buy plain old CBD oil too.

CBD comes up almost daily in my discussions with colleagues and patients. They often ask me how glutathione compares or relates to what CBD does in the body. The jury is still out on whether the CBD craze is just a fad or will have staying power, but as research has begun to mount, my guess is that the latter will be true. I'm especially encouraged by findings in the area of epilepsy in children (the FDA recently approved the first CBD-related epilepsy drug for children) as well as neurological and anxiety disorders.

CBD has a lot in common with glutathione. For one thing, they're endogenous: Just as our body produces glutathione on its own, it also produces endocannabinoids—natural cannabis-like signaling molecules that help us maintain equilibrium. And like glutathione, CBD is involved in a wide range of physiological processes. We have receptors for endocannabinoids almost everywhere (skin, bone, liver, heart, gastrointestinal tract, and so on), which may be why they influence so many aspects of our life, including memory, mood, appetite, metabolism, sleep, pain, and immunity.

Up until recently, CBD was one of those things to which I took a wait-and-see approach, in part because I wanted to watch what happened with the federal regulations governing cannabis. But what most people don't know is that CBD doesn't have to come from the cannabis plant; there are also several other noncannabis plants from which to extract CBD. (CBD also has no THC, the chemical responsible for cannabis's psychoactive effects.) As a matter of fact, cutting-edge researchers in this field have already been extracting CBD from algae plants, and are noting its high bioavailability.

Recently, Dr. Tran and I have been in the lab doing research on how water-soluble glutathione and water-soluble CBD molecules (extracted from a noncannabis hemp plant) function together. After both molecules went through separate patented technologies to create water-soluble and subnano forms, we combined them in a solution.

We speculated that if glutathione was the master of all cellular function and CBD aided in most nervous system functions, putting them together could be a one-two punch that would help the whole body function at optimal levels. Our results have been very compelling, and I've been pleasantly surprised at how well the glutathione and CBD play together. The GSH and CBD molecules seem to be compatible, even though glutathione is usually highly reactive to most other substances. Maybe it's because they both have the same inherent mission: to help the body fight off the bad guys as well as heal itself.

———

When I started my career as a pharmacist, I never in a million years thought my legacy would be being the "glutathione guy." But outside of my humanitarian efforts, focusing on GSH and other substances our bodies make themselves is my passion. The human body is indeed a wonderland, and I consider it the greatest blessing to be able to contribute even small pieces of the puzzle that is optimal human health. I pray for the ability to continue adding to the story of wellness.

And I hope that all that you have learned about glutathione in these pages will spark a revolution in you. Take action, change your life, do all the things that will ensure that you look and feel great for many years to come!

More questions about glutathione? Here's a resource for you: www.GlutathioneRevolution.com

ACTION PLAN AGENDAS

W HILE I LAID OUT THE DETAILS OF EACH ACTION PLAN EARLIER IN THE book, these at-a-glance charts pull all the steps together in an easy-to-follow format. They'll help you see exactly what you need to do and when you need to do it. Only caveat: Be sure to read the detailed plans first, so you don't miss out on any of the finer points.

ACTION PLAN ONE
KICK YOUR GSH-MAKING MACHINERY INTO GEAR

	MEALS	SUPPLEMENTATION
Sunday	Breakfast—Get-the-Glow Protein Shake Snack—¼ cup Nutty Raw Mix Lunch—1 Super Fruit Bowl; Field of Greens Salad; 1 date Dinner—Ginger-Garlic Vegetable Stir-Fry over Rice	Morning—300 mg coenzyme Q10; 100 mcg selenium (if not eating Brazil nuts); 1,000 mg vitamin C With dinner—650 mg betaine HCl + 1 capsule of digestive enzyme Evening—100–600 mg magnesium glycinate; 400 mg mixed tocopherols vitamin E
Monday	Breakfast—Get-the-Glow Protein Shake Snack—¼ cup Nutty Raw Mix Lunch—1 Super Fruit Bowl; Hearty Avo-Veg Sandwich; 1 date Dinner—Garlic and Thyme Roasted Vegetables	Morning—300 mg coenzyme Q10; 100 mcg selenium (if not eating Brazil nuts); 1,000 mg vitamin C With dinner—650 mg betaine HCl + 1 capsule of digestive enzyme Evening—100–600 mg magnesium glycinate; 400 mg mixed tocopherols vitamin E
Tuesday	Breakfast—Get-the-Glow Protein Shake Snack—¼ cup Nutty Raw Mix Lunch—1 Super Fruit Bowl; Field of Greens Salad; 1 date Dinner—Savory Mixed Bean Soup	Morning—300 mg coenzyme Q10; 100 mcg selenium (if not eating Brazil nuts); 1,000 mg vitamin C With dinner—650 mg betaine HCl + 1 capsule of digestive enzyme Evening—100–600 mg magnesium glycinate; 400 mg mixed tocopherols vitamin E
Wednesday	Breakfast—Get-the-Glow Protein Shake Snack—¼ cup Nutty Raw Mix Lunch—1 Super Fruit Bowl; Hearty Avo-Veg Sandwich; 1 date Dinner—Steamed Vegetables with Guacamole	Morning—300 mg coenzyme Q10; 100 mcg selenium (if not eating Brazil nuts); 1,000 mg vitamin C With dinner—650 mg betaine HCl + 1 capsule of digestive enzyme Evening—100–600 mg magnesium glycinate; 400 mg mixed tocopherols vitamin E

EXERCISE	DE-STRESS
Leisurely walking, 30 to 60 minutes	
Cardio workout of your choice with intervals, 30 minutes (or as long as you can)	
Strength training using your own body weight, 2 to 4 sets of 10 repetitions	3- to 7-minute meditation, twice during the day
Yoga routine or all-over body stretching routine, 30 to 60 minutes	

ACTION PLAN ONE *(continued)*

	MEALS	SUPPLEMENTATION
Thursday	Breakfast—Get-the-Glow Protein Shake Snack—¼ cup Nutty Raw Mix Lunch—1 Super Fruit Bowl; Field of Greens Salad; 1 date Dinner—Stir-Fried Vegetables with Quinoa	Morning—300 mg coenzyme Q10; 100 mcg selenium (if not eating Brazil nuts); 1,000 mg vitamin C With dinner—650 mg betaine HCl + 1 capsule of digestive enzyme Evening—100–600 mg magnesium glycinate; 400 mg mixed tocopherols vitamin E
Friday	Breakfast—Get-the-Glow Protein Shake Snack—¼ cup Nutty Raw Mix Lunch—1 Super Fruit Bowl; Hearty Avo-Veg Sandwich; 1 date Dinner—Lentil and Vegetable Soup	Morning—300 mg coenzyme Q10; 100 mcg selenium (if not eating Brazil nuts); 1,000 mg vitamin C With dinner—650 mg betaine HCl + 1 capsule of digestive enzyme Evening—100–600 mg magnesium glycinate; 400 mg mixed tocopherols vitamin E
Saturday	Breakfast—Get-the-Glow Protein Shake Snack—¼ cup Nutty Raw Mix Lunch—1 Super Fruit Bowl; Field of Greens Salad; 1 date Dinner—Pan-Seared Fish with Raspberry Vinaigrette	Morning—300 mg coenzyme Q10; 100 mcg selenium (if not eating Brazil nuts); 1,000 mg vitamin C With dinner—650 mg betaine HCl + 1 capsule of digestive enzyme Evening—100–600 mg magnesium glycinate; 400 mg mixed tocopherols vitamin E

EXERCISE	DE-STRESS
Cardio workout of your choice with intervals, 30 minutes (or as long as you can)	
Strength training using your own body weight, 2 to 4 sets of 10 repetitions	
Yoga routine or all-over body stretching routine, 30 to 60 minutes	3- to 7-minute meditation, twice during the day

ACTION PLAN TWO
DETOXING WITH GLUTATHIONE / LIGHT DETOX
Repeat everything the following week for a total of 14 days.

MEAL	NOURISH YOUR LIVER	SUPPLEMENTATION
Sunday — Breakfast—Get-the-Glow Protein Shake; Snack—¼ cup Nutty Raw Mix; Lunch—1 Super Fruit Bowl; Field of Greens Salad; 1 date; Dinner—Ginger-Garlic Vegetable Stir-Fry over Rice	Add ½ teaspoon of turmeric to your protein shake and dandelion greens or Brussels sprouts to your stir-fry. Alternatively, take a milk thistle (silymarin) or Cordyceps supplement (take these supplements according to instructions on the label).	Morning—300 mg coenzyme Q10; 100 mcg selenium (if not eating Brazil nuts); 1,000 mg vitamin C. With dinner—650 mg betaine HCl + 1 capsule of digestive enzyme. Evening—100–600 mg magnesium glycinate; 400 mg mixed tocopherols vitamin E
Monday — Breakfast—Get-the-Glow Protein Shake; Snack—¼ cup Nutty Raw Mix; Lunch—1 Super Fruit Bowl; Hearty Avo-Veg Sandwich; 1 date; Dinner—Garlic and Thyme Roasted Vegetables	Add ½ teaspoon of turmeric to your protein shake. Your dinner already has liver-boosting Brussels sprouts, so nothing extra needed today. If you're taking milk thistle (silymarin) or Cordyceps supplement, continue as directed on supplement labels.	Morning—300 mg coenzyme Q10; 100 mcg selenium (if not eating Brazil nuts); 1,000 mg vitamin C. With dinner—650 mg betaine HCl + 1 capsule of digestive enzyme. Evening—100–600 mg magnesium glycinate; 400 mg mixed tocopherols vitamin E
Tuesday — Breakfast—Get-the-Glow Protein Shake; Snack—¼ cup Nutty Raw Mix; Lunch—1 Super Fruit Bowl; Field of Greens Salad; 1 date; Dinner—Savory Mixed Bean Soup	Add ½ teaspoon of turmeric to your protein shake. Add dandelion greens to your soup, stirring in the last minute until wilted. If you're taking milk thistle (silymarin) or Cordyceps supplement, continue as directed on supplement labels.	Morning—300 mg coenzyme Q10; 100 mcg selenium (if not eating Brazil nuts); 1,000 mg vitamin C. With dinner—650 mg betaine HCl + 1 capsule of digestive enzyme. Evening—100–600 mg magnesium glycinate; 400 mg mixed tocopherols vitamin E

GLUTATHIONE	EXERCISE	DE-STRESS	UNPLUG
Once a day: 0.5 ml (that's 4 sprays) if you're using Auro GSH; follow instructions on the label when using other sources of glutathione	Leisurely walking, 30 to 60 minutes		Take at least an hour a day away from electronics. Charge phones, electronic pads, and computers in a room other than your bedroom
Once a day: 0.5 ml (that's 4 sprays) if you're using Auro GSH; follow instructions on the label when using other sources of glutathione	Cardio workout of your choice with intervals, 30 minutes (or as long as you can)		Take at least an hour a day away from electronics. Charge phones, electronic pads, and computers in a room other than your bedroom
Once a day: 0.5 ml (that's 4 sprays) if you're using Auro GSH; follow instructions on the label when using other sources of glutathione	Strength training using your own body weight, 2 to 4 sets of 10 repetitions	3- to 7-minute meditation, twice during the day	Take at least an hour a day away from electronics. Charge phones, electronic pads, and computers in a room other than your bedroom

ACTION PLAN TWO—LIGHT DETOX *(continued)*

	MEAL	NOURISH YOUR LIVER	SUPPLEMENTATION
Wednesday	Breakfast—Get-the-Glow Protein Shake Snack—¼ cup Nutty Raw Mix Lunch—1 Super Fruit Bowl; Hearty Avo-Veg Sandwich; 1 date Dinner—Steamed Vegetables with Guacamole	Add ½ teaspoon of turmeric to your protein shake. Steam an artichoke (trim leaves, place stem up in a steamer basket, and steam for 20 to 30 minutes). Serve plain or with vinaigrette (page 229) as a starter alongside your Steamed Vegetables with Guacamole dinner. If you're taking milk thistle (silymarin) or Cordyceps supplement, continue as directed on supplement labels.	Morning—300 mg coenzyme Q10; 100 mcg selenium (if not eating Brazil nuts); 1,000 mg vitamin C With dinner—650 mg betaine HCl + 1 capsule of digestive enzyme Evening—100–600 mg magnesium glycinate; 400 mg mixed tocopherols vitamin E
Thursday	Breakfast—Get-the-Glow Protein Shake Snack—¼ cup Nutty Raw Mix Lunch—1 Super Fruit Bowl; Field of Greens Salad; 1 date Dinner—Stir-Fried Vegetables with Quinoa	Add ½ teaspoon of turmeric to your protein shake, and dandelion greens or Brussels sprouts to your stir-fry. If you're taking milk thistle (silymarin) or Cordyceps supplement, continue as directed on supplement labels.	Morning—300 mg coenzyme Q10; 100 mcg selenium (if not eating Brazil nuts); 1,000 mg vitamin C With dinner—650 mg betaine HCl + 1 capsule of digestive enzyme Evening—100–600 mg magnesium glycinate; 400 mg mixed tocopherols vitamin E
Friday	Breakfast—Get-the-Glow Protein Shake Snack—¼ cup Nutty Raw Mix Lunch—1 Super Fruit Bowl; Hearty Avo-Veg Sandwich; 1 date Dinner—Lentil and Vegetable Soup	Add ½ teaspoon of turmeric to your protein shake, and dandelion greens to your soup, stirring in the last minute until wilted. If you're taking milk thistle (silymarin) or Cordyceps supplement, continue as directed on supplement labels.	Morning—300 mg coenzyme Q10; 100 mcg selenium (if not eating Brazil nuts); 1,000 mg vitamin C With dinner—650 mg betaine HCl + 1 capsule of digestive enzyme Evening—100–600 mg magnesium glycinate; 400 mg mixed tocopherols vitamin E

GLUTATHIONE	EXERCISE	DE-STRESS	UNPLUG
Once a day: 0.5 ml (that's 4 sprays) if you're using Auro GSH; follow instructions on the label when using other sources of glutathione	Yoga routine or all-over body stretching routine, 30 to 60 minutes		Take at least an hour a day away from electronics. Charge phones, electronic pads, and computers in a room other than your bedroom
Once a day: 0.5 ml (that's 4 sprays) if you're using Auro GSH; follow instructions on the label when using other sources of glutathione	Cardio workout of your choice with intervals, 30 minutes (or as long as you can)		Take at least an hour a day away from electronics. Charge phones, electronic pads, and computers in a room other than your bedroom
Once a day: 0.5 ml (that's 4 sprays) if you're using Auro GSH; follow instructions on the label when using other sources of glutathione	Strength training using your own body weight, 2 to 4 sets of 10 repetitions		Take at least an hour a day away from electronics. Charge phones, electronic pads, and computers in a room other than your bedroom

ACTION PLAN TWO—LIGHT DETOX *(continued)*

	MEAL	NOURISH YOUR LIVER	SUPPLEMENTATION
Saturday	Breakfast—Get-the-Glow Protein Shake Snack—¼ cup Nutty Raw Mix Lunch—1 Super Fruit Bowl; Field of Greens Salad; 1 date Dinner—Pan-Seared Fish with Raspberry Vinaigrette	Add ½ teaspoon turmeric to your protein shake. Steam an artichoke (trim leaves, place in a steamer basket, stem up, and steam for 20 to 30 minutes). Serve plain or with vinaigrette (page 229) as a starter alongside your Pan-Seared Fish dinner. If you're taking milk thistle (silymarin) or Cordyceps supplement, continue as directed on supplement labels.	Morning—300 mg coenzyme Q10; 100 mcg selenium (if not eating Brazil nuts); 1,000 mg vitamin C With dinner—650 mg betaine HCl + 1 capsule of digestive enzyme Evening—100–600 mg magnesium glycinate; 400 mg mixed tocopherols vitamin E

GLUTATHIONE	EXERCISE	DE-STRESS	UNPLUG
Once a day: 0.5 ml (that's 4 sprays) if you're using Auro GSH; follow instructions on the label when using other sources of glutathione	Yoga routine or all-over body stretching routine, 30 to 60 minutes	3- to 7-minute meditation, twice during the day	Take at least an hour a day away from electronics. Charge phones, electronic pads, and computers in a room other than your bedroom

ACTION PLAN TWO
DETOXING WITH GLUTATHIONE / HEAVY DETOX

Repeat everything the following week for a total of 14 days.

	MEAL	NOURISH YOUR LIVER	SUPPLEMENTATION
Sunday	Breakfast—Get-the-Glow Protein Shake Snack—¼ cup Nutty Raw Mix Lunch—1 Super Fruit Bowl; Field of Greens Salad; 1 date Dinner—Ginger-Garlic Vegetable Stir-Fry over Rice	Add ½ teaspoon of turmeric to your protein shake and dandelion greens or Brussels sprouts to your stir-fry. Alternatively, take a milk thistle (silymarin) or Cordyceps supplement (take these supplements according to instructions on the label).	Morning—300 mg coenzyme Q10; 100 mcg selenium (if not eating Brazil nuts); 1,000 mg vitamin C; 200 mcg chromium picolinate (discontinue after 9 to 12 months); NAC; optional: 600 mg alpha lipoic acid; B-complex With dinner—650 mg betaine HCl + 1 capsule of digestive enzyme Evening—100–600 mg magnesium glycinate; 400 mg mixed tocopherols vitamin E; 200 mcg chromium picolinate (discontinue after 9 to 12 months); 30 mg copper & 2 mg zinc
Monday	Breakfast—Get-the-Glow Protein Shake Snack—¼ cup Nutty Raw Mix Lunch—1 Super Fruit Bowl; Hearty Avo-Veg Sandwich; 1 date Dinner—Garlic and Thyme Roasted Vegetables	Add ½ teaspoon of turmeric to your protein shake. Your dinner already has liver-boosting Brussels sprouts, so nothing extra needed today. If you're taking milk thistle (silymarin) or Cordyceps supplement, continue as directed on supplement labels.	Morning—300 mg coenzyme Q10; 100 mcg selenium (if not eating Brazil nuts); 1,000 mg vitamin C; 200 mcg chromium picolinate (discontinue after 9 to 12 months); NAC; optional: 600 mg alpha lipoic acid; B-complex With dinner—650 mg betaine HCl + 1 capsule of digestive enzyme Evening—100–600 mg magnesium glycinate; 400 mg mixed tocopherols vitamin E; 200 mcg chromium picolinate (discontinue after 9 to 12 months); 30 mg copper & 2 mg zinc

GLUTATHIONE	EXERCISE	DE-STRESS	UNPLUG
Once a day: 0.5 ml (that's 4 sprays) if you're using Auro GSH; follow instructions on the label when using other sources of glutathione	Leisurely walking, 30 to 60 minutes		Take at least an hour a day away from electronics. Charge phones, electronic pads, and computers in a room other than your bedroom
Once a day: 0.5 ml (that's 4 sprays) if you're using Auro GSH; follow instructions on the label when using other sources of glutathione	Cardio workout of your choice with intervals, 30 minutes (or as long as you can)		Take at least an hour a day away from electronics. Charge phones, electronic pads, and computers in a room other than your bedroom

ACTION PLAN TWO—HEAVY DETOX *(continued)*

MEAL	NOURISH YOUR LIVER	SUPPLEMENTATION
Tuesday Breakfast—Get-the-Glow Protein Shake Snack—¼ cup Nutty Raw Mix Lunch—1 Super Fruit Bowl; Field of Greens Salad; 1 date Dinner—Savory Mixed Bean Soup	Add ½ teaspoon of turmeric to your protein shake. Add dandelion greens to your soup, stirring in the last minute until wilted. If you're taking milk thistle (silymarin) or Cordyceps supplement, continue as directed on supplement labels.	Morning—300 mg coenzyme Q10; 100 mcg selenium (if not eating Brazil nuts); 1,000 mg vitamin C; 200 mcg chromium picolinate (discontinue after 9 to 12 months); NAC; optional: 600 mg alpha lipoic acid; B-complex With dinner—650 mg betaine HCl + 1 capsule of digestive enzyme Evening—100–600 mg magnesium glycinate; 400 mg mixed tocopherols vitamin E; 200 mcg chromium picolinate (discontinue after 9 to 12 months); 30 mg copper & 2 mg zinc
Wednesday Breakfast—Get-the-Glow Protein Shake Snack—¼ cup Nutty Raw Mix Lunch—1 Super Fruit Bowl; Hearty Avo-Veg Sandwich; 1 date Dinner—Steamed Vegetables with Guacamole	Add ½ teaspoon of turmeric to your protein shake. Steam an artichoke (trim leaves, place stem up in a steamer basket, and steam for 20 to 30 minutes. Serve plain or with vinaigrette (page 229) as a starter alongside your Steamed Vegetables with Guacamole dinner. If you're taking milk thistle (silymarin) or Cordyceps supplement, continue as directed on supplement labels.	Morning—300 mg coenzyme Q10; 100 mcg selenium (if not eating Brazil nuts); 1,000 mg vitamin C; 200 mcg chromium picolinate (discontinue after 9 to 12 months); NAC; optional: 600 mg alpha lipoic acid; B-complex With dinner—650 mg betaine HCl + 1 capsule of digestive enzyme Evening—100–600 mg magnesium glycinate; 400 mg mixed tocopherols vitamin E; 200 mcg chromium picolinate (discontinue after 9 to 12 months); 30 mg copper & 2 mg zinc

GLUTATHIONE	EXERCISE	DE-STRESS	UNPLUG
Once a day: 0.5 ml (that's 4 sprays) if you're using Auro GSH; follow instructions on the label when using other sources of glutathione	Strength training using your own body weight, 2 to 4 sets of 10 repetitions	3- to 7-minute meditation, twice during the day	Take at least an hour a day away from electronics. Charge phones, electronic pads, and computers in a room other than your bedroom
Once a day: 0.5 ml (that's 4 sprays) if you're using Auro GSH; follow instructions on the label when using other sources of glutathione	Yoga routine or all-over body stretching routine, 30 to 60 minutes		Take at least an hour a day away from electronics. Charge phones, electronic pads, and computers in a room other than your bedroom

ACTION PLAN TWO—HEAVY DETOX *(continued)*

MEAL	NOURISH YOUR LIVER	SUPPLEMENTATION
Thursday Breakfast—Get-the-Glow Protein Shake Snack—¼ cup Nutty Raw Mix Lunch—1 Super Fruit Bowl; Field of Greens Salad; 1 date Dinner—Stir-Fried Vegetables with Quinoa	Add ½ teaspoon of turmeric to your protein shake, and dandelion greens or Brussels sprouts to your stir-fry. If you're taking milk thistle (silymarin) or Cordyceps supplement, continue as directed on supplement labels.	Morning—300 mg coenzyme Q10; 100 mcg selenium (if not eating Brazil nuts); 1,000 mg vitamin C; 200 mcg chromium picolinate (discontinue after 9 to 12 months); NAC; optional: 600 mg alpha lipoic acid; B-complex With dinner—650 mg betaine HCl + 1 capsule of digestive enzyme Evening—100–600 mg magnesium glycinate; 400 mg mixed tocopherols vitamin E; 200 mcg chromium picolinate (discontinue after 9 to 12 months); 30 mg copper & 2 mg zinc
Friday Breakfast—Get-the-Glow Protein Shake Snack—¼ cup Nutty Raw Mix Lunch—1 Super Fruit Bowl; Hearty Avo-Veg Sandwich; 1 date Dinner—Lentil and Vegetable Soup	Add ½ teaspoon of turmeric to your protein shake, and dandelion greens to your soup, stirring in the last minute until wilted. If you're taking milk thistle (silymarin) or Cordyceps supplement, continue as directed on supplement labels.	Morning—300 mg coenzyme Q10; 100 mcg selenium (if not eating Brazil nuts); 1,000 mg vitamin C; 200 mcg chromium picolinate (discontinue after 9 to 12 months); NAC; optional: 600 mg alpha lipoic acid; B-complex With dinner—650 mg betaine HCl + 1 capsule of digestive enzyme Evening—100–600 mg magnesium glycinate; 400 mg mixed tocopherols vitamin E; 200 mcg chromium picolinate (discontinue after 9 to 12 months); 30 mg copper & 2 mg zinc

GLUTATHIONE	EXERCISE	DE-STRESS	UNPLUG
Once a day: 0.5 ml (that's 4 sprays) if you're using Auro GSH; follow instructions on the label when using other sources of glutathione	Cardio workout of your choice with intervals, 30 minutes (or as long as you can)		Take at least an hour a day away from electronics. Charge phones, electronic pads, and computers in a room other than your bedroom
Once a day: 0.5 ml (that's 4 sprays) if you're using Auro GSH; follow instructions on the label when using other sources of glutathione	Strength training using your own body weight, 2 to 4 sets of 10 repetitions		Take at least an hour a day away from electronics. Charge phones, electronic pads, and computers in a room other than your bedroom

ACTION PLAN TWO—HEAVY DETOX *(continued)*

MEAL	NOURISH YOUR LIVER	SUPPLEMENTATION
Saturday Breakfast—Get-the-Glow Protein Shake Snack—¼ cup Nutty Raw Mix Lunch—1 Super Fruit Bowl; Field of Greens Salad; 1 date Dinner—Pan-Seared Fish with Raspberry Vinaigrette	Add ½ teaspoon turmeric to your protein shake. Steam an artichoke (trim leaves, place in a steamer basket, stem up, and steam for 20 to 30 minutes). Serve plain or with vinaigrette (page 229) as a starter alongside your Pan-Seared Fish dinner. If you're taking milk thistle (silymarin) or Cordyceps supplement, continue as directed on supplement labels.	Morning—300 mg coenzyme Q10; 100 mcg selenium (if not eating Brazil nuts); 1,000 mg vitamin C; 200 mcg chromium picolinate (discontinue after 9 to 12 months); NAC; optional: 600 mg alpha lipoic acid; B-complex With dinner—650 mg betaine HCl + 1 capsule of digestive enzyme Evening—100–600 mg magnesium glycinate; 400 mg mixed tocopherols vitamin E; 200 mcg chromium picolinate (discontinue after 9 to 12 months); 30 mg copper & 2 mg zinc

GLUTATHIONE	EXERCISE	DE-STRESS	UNPLUG
Once a day: 0.5 ml (that's 4 sprays) if you're using Auro GSH; follow instructions on the label when using other sources of glutathione	Yoga routine or all-over body stretching routine, 30 to 60 minutes	3- to 7-minute meditation, twice during the day	Take at least an hour a day away from electronics. Charge phones, electronic pads, and computers in a room other than your bedroom

ACTION PLAN THREE
GSH FOR BETTER, BRIGHTER SKIN
Repeat everything the following week for a total of 14 days.

	MEAL	NOURISH YOUR LIVER	SUPPLEMENTATION
Sunday	Breakfast—Get-the-Glow Protein Shake Snack—¼ cup Nutty Raw Mix Lunch—1 Super Fruit Bowl; Field of Greens Salad; 1 date Dinner—Ginger-Garlic Vegetable Stir-Fry over Rice	Add ½ teaspoon of turmeric to your protein shake and dandelion greens or Brussels sprouts to your stir-fry. Alternatively, take a milk thistle (silymarin) or Cordyceps supplement (take these supplements according to instructions on the label).	Morning—300 mg coenzyme Q10; 100 mcg selenium (if not eating Brazil nuts); 1,000 mg vitamin C With dinner—650 mg betaine HCl + 1 capsule of digestive enzyme Evening—100–600 mg magnesium glycinate; 400 mg mixed tocopherols vitamin E
Monday	Breakfast—Get-the-Glow Protein Shake Snack—¼ cup Nutty Raw Mix Lunch—1 Super Fruit Bowl; Hearty Avo-Veg Sandwich; 1 date Dinner—Garlic and Thyme Roasted Vegetables	Add ½ teaspoon of turmeric to your protein shake. Your dinner already has liver-boosting Brussels sprouts, so nothing extra needed today. If you're taking milk thistle (silymarin) or Cordyceps supplement, continue as directed on supplement labels.	Morning—300 mg coenzyme Q10; 100 mcg selenium (if not eating Brazil nuts); 1,000 mg vitamin C With dinner—650 mg betaine HCl + 1 capsule of digestive enzyme Evening—100–600 mg magnesium glycinate; 400 mg mixed tocopherols vitamin E
Tuesday	Breakfast—Get-the-Glow Protein Shake Snack—¼ cup Nutty Raw Mix Lunch—1 Super Fruit Bowl; Field of Greens Salad; 1 date Dinner—Savory Mixed Bean Soup	Add ½ teaspoon of turmeric to your protein shake. Add dandelion greens to your soup, stirring in the last minute until wilted. If you're taking milk thistle (silymarin) or Cordyceps supplement, continue as directed on supplement labels.	Morning—300 mg coenzyme Q10; 100 mcg selenium (if not eating Brazil nuts); 1,000 mg vitamin C With dinner—650 mg betaine HCl + 1 capsule of digestive enzyme Evening—100–600 mg magnesium glycinate; 400 mg mixed tocopherols vitamin E

GLUTATHIONE	EXERCISE	DE-STRESS	UNPLUG	SKIN CARE
Once a day: 0.5 ml (that's 4 sprays) if you're using Auro GSH; follow instructions on the label when using other sources of glutathione	Leisurely walking, 30 to 60 minutes		Take at least an hour a day away from electronics. Charge phones, electronic pads, and computers in a room other than your bedroom	<u>Morning</u> Cleanse skin; treat with topical glutathione (if not following Action Plan Two/Light Detox); moisturize and apply sunscreen <u>Evening</u> Exfoliate skin; treat with any antiaging serums or moisturizers (personal choice)
Once a day: 0.5 ml (that's 4 sprays) if you're using Auro GSH; follow instructions on the label when using other sources of glutathione	Cardio workout of your choice with intervals, 30 minutes (or as long as you can)		Take at least an hour a day away from electronics. Charge phones, electronic pads, and computers in a room other than your bedroom	<u>Morning</u> Cleanse skin; treat with topical glutathione (if not following Action Plan Two/Light Detox); moisturize and apply sunscreen <u>Evening</u> Exfoliate skin; treat with any antiaging serums or moisturizers (personal choice)
Once a day: 0.5 ml (that's 4 sprays) if you're using Auro GSH; follow instructions on the label when using other sources of glutathione	Strength training using your own body weight, 2 to 4 sets of 10 repetitions	3- to 7-minute meditation, twice during the day	Take at least an hour a day away from electronics. Charge phones, electronic pads, and computers in a room other than your bedroom	<u>Morning</u> Cleanse skin; treat with topical glutathione (if not following Action Plan Two/Light Detox); moisturize and apply sunscreen <u>Evening</u> Exfoliate skin; treat with any antiaging serums or moisturizers (personal choice)

ACTION PLAN THREE *(continued)*

	MEAL	NOURISH YOUR LIVER	SUPPLEMENTATION
Wednesday	Breakfast—Get-the-Glow Protein Shake Snack—¼ cup Nutty Raw Mix Lunch—1 Super Fruit Bowl; Hearty Avo-Veg Sandwich; 1 date Dinner—Steamed Vegetables with Guacamole	Add ½ teaspoon of turmeric to your protein shake. Steam an artichoke (trim leaves, place stem up in a steamer basket, and steam for 20 to 30 minutes). Serve plain or with vinaigrette (page 229) as a starter alongside your Steamed Vegetables with Guacamole dinner. If you're taking milk thistle (silymarin) or Cordyceps supplement, continue as directed on supplement labels.	Morning—300 mg coenzyme Q10; 100 mcg selenium (if not eating Brazil nuts); 1,000 mg vitamin C With dinner—650 mg betaine HCl + 1 capsule of digestive enzyme Evening—100–600 mg magnesium glycinate; 400 mg mixed tocopherols vitamin E
Thursday	Breakfast—Get-the-Glow Protein Shake Snack—¼ cup Nutty Raw Mix Lunch—1 Super Fruit Bowl; Field of Greens Salad; 1 date Dinner—Stir-Fried Vegetables with Quinoa	Add ½ teaspoon of turmeric to your protein shake, and dandelion greens or Brussels sprouts to your stir-fry. If you're taking milk thistle (silymarin) or Cordyceps supplement, continue as directed on supplement labels.	Morning—300 mg coenzyme Q10; 100 mcg selenium (if not eating Brazil nuts); 1,000 mg vitamin C With dinner—650 mg betaine HCl + 1 capsule of digestive enzyme Evening—100–600 mg magnesium glycinate; 400 mg mixed tocopherols vitamin E
Friday	Breakfast—Get-the-Glow Protein Shake Snack—¼ cup Nutty Raw Mix Lunch—1 Super Fruit Bowl; Hearty Avo-Veg Sandwich; 1 date Dinner—Lentil and Vegetable Soup	Add ½ teaspoon of turmeric to your protein shake, and dandelion greens to your soup, stirring in the last minute until wilted. If you're taking milk thistle (silymarin) or Cordyceps supplement, continue as directed on supplement labels.	Morning—300 mg coenzyme Q10; 100 mcg selenium (if not eating Brazil nuts); 1,000 mg vitamin C With dinner—650 mg betaine HCl + 1 capsule of digestive enzyme Evening—100–600 mg magnesium glycinate; 400 mg mixed tocopherols vitamin E

GLUTATHIONE	EXERCISE	DE-STRESS	UNPLUG	SKIN CARE
Once a day: 0.5 ml (that's 4 sprays) if you're using Auro GSH; follow instructions on the label when using other sources of glutathione	Yoga routine or all-over body stretching routine, 30 to 60 minutes		Take at least an hour a day away from electronics. Charge phones, electronic pads, and computers in a room other than your bedroom	**Morning** Cleanse skin; treat with topical glutathione (if not following Action Plan Two/Light Detox); moisturize and apply sunscreen **Evening** Exfoliate skin; treat with any antiaging serums or moisturizers (personal choice)
Once a day: 0.5 ml (that's 4 sprays) if you're using Auro GSH; follow instructions on the label when using other sources of glutathione	Cardio workout of your choice with intervals, 30 minutes (or as long as you can)		Take at least an hour a day away from electronics. Charge phones, electronic pads, and computers in a room other than your bedroom	**Morning** Cleanse skin; treat with topical glutathione (if not following Action Plan Two/Light Detox); moisturize and apply sunscreen **Evening** Exfoliate skin; treat with any antiaging serums or moisturizers (personal choice)
Once a day: 0.5 ml (that's 4 sprays) if you're using Auro GSH; follow instructions on the label when using other sources of glutathione	Strength training using your own body weight, 2 to 4 sets of 10 repetitions		Take at least an hour a day away from electronics. Charge phones, electronic pads, and computers in a room other than your bedroom	**Morning** Cleanse skin; treat with topical glutathione (if not following Action Plan Two/Light Detox); moisturize and apply sunscreen **Evening** Exfoliate skin; treat with any antiaging serums or moisturizers (personal choice)

ACTION PLAN THREE (continued)

	MEAL	NOURISH YOUR LIVER	SUPPLEMENTATION
Saturday	Breakfast—Get-the-Glow Protein Shake Snack—¼ cup Nutty Raw Mix Lunch—1 Super Fruit Bowl; Field of Greens Salad; 1 date Dinner—Pan-Seared Fish with Raspberry Vinaigrette	Add ½ teaspoon turmeric to your protein shake. Steam an artichoke (trim leaves, place in a steamer basket, stem up, and steam for 20 to 30 minutes). Serve plain or with vinaigrette (page 229) as a starter alongside your Pan-Seared Fish dinner. If you're taking milk thistle (silymarin) or Cordyceps supplement, continue as directed on supplement labels.	Morning—300 mg coenzyme Q10; 100 mcg selenium (if not eating Brazil nuts); 1,000 mg vitamin C With dinner—650 mg betaine HCl + 1 capsule of digestive enzyme Evening—100–600 mg magnesium glycinate; 400 mg mixed tocopherols vitamin E

GLUTATHIONE	EXERCISE	DE-STRESS	UNPLUG	SKIN CARE
Once a day: 0.5 ml (that's 4 sprays) if you're using Auro GSH; follow instructions on the label when using other sources of glutathione	Yoga routine or all-over body stretching routine, 30 to 60 minutes	3- to 7-minute meditation, twice during the day	Take at least an hour a day away from electronics. Charge phones, electronic pads, and computers in a room other than your bedroom	Morning Cleanse skin; treat with topical glutathione (if not following Action Plan Two/Light Detox); moisturize and apply sunscreen Evening Exfoliate skin; treat with any antiaging serums or moisturizers (personal choice)

RECIPES

BREAKFAST

• Get-the-Glow Protein Shake •

Makes 1 shake

This morning shake is both filling and filled with nutrients that either increase or support glutathione production. The combination of fresh fruit and powdered greens will increase your overall daily antioxidant intake, especially of vitamins C and A. Vary the berries, if you like, to add variety to your week, and play around with the amount of water depending on how concentrated you like your shakes.

You can also use canned, unsweetened (full-fat) coconut milk instead of water, which will add flavor and make the shake feel more substantial. The flavor is very mild—it doesn't really taste "coconutty" as most coconut oils and other coconut products do. Coconut milk can also be added to coffee and anything else that you would normally put milk into (even my kids love it in their whole-grain cereal). You'll find it in cans in your local grocery store, usually in the Asian food or cooking/baking sections. Just check the ingredients and make sure that there is no added sugar, and that it is free of all other unnecessary additives. I'm partial to the Simply Organic brand. If you're coconut-averse, use your favorite unsweetened nut milk instead.

15 grams (approx. 2 scoops) nondenatured whey isolate protein powder, such as Ultra Meal by Metagenics, or plant-based protein powder

5 grams soluble/insoluble fiber powder, such as Ultra Fiber Plus by Medifood

8 grams (approx. 1 scoop) powdered greens or powdered reds, such as Nano Greens or Nano Reds

½ cup fresh or frozen blueberries, raspberries, or a combination of the two

8 to 12 ounces filtered water or shaken, full-fat, unsweetened coconut milk

1 to 2 pitted dates for sweetness (optional)

In a blender, combine all the ingredients. Blend until smooth.

SNACKS

· Nutty Raw Mix ·

Makes about 10 ¼-cup servings

Walnuts and pecans are particularly high in anti-oxidants, and Brazil nuts—the most important nut in the bunch—are a superrich source of selenium. This mix is designed to maximize the nutrients you need for glutathione-building, but if there's one particular ingredient you don't like, just leave it out and add more of another nut (except the Brazil nuts). Note that buying raw nuts in bulk can save you money. Store leftover nuts in the freezer—otherwise, because of their high oil content, they can go bad after a few weeks.

½ cup raw walnuts
40 raw Brazil nuts
½ cup raw cashews
½ cup raw pecans
½ cup raw almonds
¼ cup raisins (optional)

Combine the nuts and raisins, if using, in an airtight container. Store in a cool, dark place or in the refrigerator. When you dish out ¼-cup servings, only include four Brazil nuts in each.

LUNCH

· Super Fruit Bowl ·

Makes 1 serving

Toss an orange or banana into your lunch box and it can be, well, sort of boring. To me at least, there's something a little more enticing about a bowl of cut-up fruit, so I take that extra step when making my own lunch. The bowl suggestions (and they are just suggestions) below are combinations of fruits that are high either in glutathione, cysteine, vitamin C, or all of the above. Choose one pairing, or you are welcome to arrange them any way you like and to increase the quantities if appetite demands.

½ cup cubed papaya plus 1 banana, sliced

½ cup cubed cantaloupe plus ½ cup cubed watermelon

½ orange or 1 tangerine, sectioned, plus ½ grapefruit, sectioned

1 banana, sliced, plus 1 peach, sliced

½ cup sliced strawberries plus ½ cup cubed papaya

½ cup cubed cantaloupe plus ½ grapefruit, sectioned

½ cup cubed watermelon plus 1 peach, sliced

Mix your preferred combination of fruit in a bowl.

· Field of Greens Salad ·

Makes 1 serving

This simple salad contains two good sources of both glutathione and vitamin C: tomato and red pepper. The avocado is added for both its glutathione and its healthy fat content. I've included ideas for salad dressing, all easy to make. But if you prefer store-bought, make sure you buy a natural dressing without added chemical preservatives.

2 to 3 cups mixed greens (spring mix or baby lettuce is good)

¼ cup sliced cucumber

1 carrot, sliced

½ avocado, peeled, pitted, then sliced or diced

½ tomato, sliced

¼ red bell pepper, seeded and diced

2 tablespoons Dijon Vinaigrette Dressing (recipe follows), Raspberry Vinaigrette (page 229), or 1½ teaspoons extra-virgin olive oil plus 1 teaspoon balsamic vinegar

Sea salt and freshly ground black pepper (optional)

Place the greens and vegetables in a bowl. Drizzle with your choice of dressing and toss. Season with salt and pepper to taste, if desired.

Dijon Vinaigrette Dressing

2 tablespoon red wine vinegar

1 teaspoon Dijon mustard

½ garlic clove, minced

6 tablespoons extra-virgin olive oil

Sea salt and freshly ground black pepper

In a lidded jar, combine the vinegar, mustard, and garlic. Whisk in the olive oil. Season with salt and pepper to taste. Store in the refrigerator.

· Hearty Avo-Veg Sandwich ·

Makes 1 sandwich

Next to asparagus, avocado has the most glutathione of any vegetable (although it's actually a fruit, but who doesn't think of it as a veggie?). Whichever bread you choose for your sandwich, check the ingredients list to make sure it contains no sugar or corn syrup. Also make sure the condiments you choose are all natural and don't have added chemicals and fillers.

Pesto (such as Trader Joe's vegan or homemade), salad dressing (a little Thousand Island, for instance), or light mayonnaise

2 slices whole wheat, whole grain, or gluten-free bread

1 romaine lettuce leaf, rib removed

1 square slice cheese of your choice

2 to 4 thin slices tomato

6 thin slices cucumber

½ avocado, peeled, pitted, and sliced

Sliced and seeded jalapeño peppers (optional)

Sea salt and freshly ground black pepper

Spread your condiment or dressing of choice on both slices of the bread. Layer the vegetables on one slice, then season with salt and pepper to taste. Top with the remaining slice of bread and cut the sandwich in half to serve.

DINNER

· Ginger-Garlic Vegetable Stir-Fry over Rice ·
Makes 2 servings

This recipe calls for many high-glutathione vegetables, including bell peppers, broccoli, and asparagus. Feel free to add or substitute other vegetable sources of glutathione from the list on page 143.

2 tablespoons clarified butter or coconut oil

½ medium-size onion, sliced

2 carrots, peeled and sliced

1 red bell pepper, seeded and chopped

½ green bell pepper, seeded and chopped

1 teaspoon minced or finely shredded fresh ginger

3 garlic cloves, minced

½ cup chopped broccoli

1½ cups sugar snap peas

5 or 6 asparagus stalks, chopped into 2-inch pieces

2 celery stalks, sliced

1 cup bok choy

Soy sauce or Bragg Liquid Aminos (a soy sauce alternative)

Red chili paste (optional)

½ cup cooked California-grown brown rice or other whole grain (see page 149 for more about rice)

Heat a wok or large skillet over medium heat and add the clarified butter. When the butter is melted, add the onion, carrots, bell peppers, ginger, and garlic to the pan and cook for 5 minutes, or until vegetables have softened slightly. Increase the heat to high and add the broccoli, peas, asparagus, celery, and bok choy. Stirring continuously, cook for 5 to 7 minutes, or until the vegetables are tender-crisp. Season with soy sauce to taste and, if desired, add red chili paste. Serve over the brown rice.

• Garlic and Thyme Roasted Vegetables •

Makes 2 servings

Although the flavors here say "autumn," you can generally find these delicious veggies any time of year. If some of them are not available in your market, fill in the blanks with carrots, cauliflower, and/or broccoli. Serve with Field of Greens Salad (page 220) and/or cauliflower rice, brown rice, or over cooked whole grains, such as farro, spelt, or whole wheat couscous, if desired.

1 small butternut squash, peeled, seeded, and cubed

1 yam, peeled and cubed

1 parsnip, peeled and cubed

1½ to 2 cups Brussels sprouts, halved

½ red onion, chopped

6 garlic cloves, chopped

3 tablespoons fresh thyme leaves, or 3 teaspoons dried

3 tablespoons extra-virgin olive oil

Sea salt and freshly ground black pepper

Preheat the oven to 425°F.

Combine all the ingredients in a large bowl. Arrange in a single layer on a baking pan or cookie sheet and bake for 35 to 45 minutes, or until tender and golden, stirring once halfway through cooking.

· Savory Mixed Bean Soup ·

Makes 6 to 8 servings

The beans and garlic in this hearty soup will add selenium to your diet. For ease and convenience, I've suggested you use a mixed-bean soup mix. Various companies make bean soup mixes, including Goya and Bob's Red Mill, and they are available in natural food stores and supermarkets. If you want to use your own beans, try navy, cannellini, or pinto. Serve with Field of Greens Salad (page 220), if desired.

1 bag (approx. 1 pound) dried mixed-bean soup mix

2 tablespoons extra-virgin olive oil

½ medium-size yellow onion, diced

4 green onions, white and green parts, finely chopped

3 garlic cloves, minced

1 tomato, finely diced

1½ teaspoons salt

Sea salt and freshly ground black pepper

Crushed red pepper flakes (optional)

Soak the dried mixed beans in room-temperature water overnight or for at least 4 to 5 hours. Rinse the soaked beans, transfer to a slow cooker or large pot, and cover with 2½ quarts of water. In a separate pan, heat the olive oil over medium heat. Add the yellow onion and cook for 5 minutes, or until soft. Transfer the onion and remaining ingredients to the pot of beans. Cover and cook until the beans are tender (in a regular pot, anywhere from 1 to 2 hours depending on the freshness of your beans; for a slow cooker, follow the manufacturer's recommendation). Add water as needed to create the consistency you like. Season with salt and pepper to taste, and, if using, red pepper flakes before serving.

• Steamed Vegetables with Guacamole •

Makes 1 or 2 servings

On their own, steamed vegetables are healthy, but let's face it, they can be bland. I use guacamole to liven them up and add healthy fats to the dish. Serve with brown rice, if desired.

1½ cups cut cauliflower (small florets)

2 carrots, peeled and diced

1½ cups chopped sugar snap peas

1 cup chopped baby green beans

Sea salt and freshly ground black pepper

Guacamole (recipe follows)

Place a steamer basket in a pot and fill with water until it reaches just below the basket. Place all the vegetables into the steamer basket and cook for 6 to 8 minutes, or until tender-crisp. Season with salt and pepper to taste and top with a dollop of guacamole.

Guacamole

1 avocado, peeled, pitted, and diced

½ Roma tomato, chopped

2 tablespoons chopped fresh cilantro

1 teaspoon diced onion

½ teaspoon freshly squeezed lime juice

Sea salt and freshly ground black pepper

Chopped serrano pepper (optional)

Place all ingredients in a small bowl, including salt and pepper to taste. Stir to combine, mashing the avocado as you go. Adjust the seasonings to taste.

· Stir-Fried Vegetables with Quinoa ·

Makes 2 to 4 servings

This simple dish supplies plenty of vitamin C (broccoli and red pepper), protein (quinoa), and selenium (mushrooms). It's light but also leaves you feeling satisfied.

2 tablespoons coconut oil

1 to 2 garlic cloves, finely chopped

½ cup chopped broccoli

2 carrots, chopped

1 medium-size onion, chopped

1 head bok choy, chopped

1 red bell pepper, seeded and chopped

1½ cups sugar snap peas

4 to 6 ounces mushrooms (any variety), sliced

1 jalapeño pepper, seeded and chopped (optional)

3 cups spinach, chopped if large leaf

2 tablespoons cider vinegar

Sea salt and freshly ground black pepper

2 cups cooked quinoa, prepared according to
package instructions

In a large sauté pan or wok, melt the coconut oil over medium-high heat. Add all the vegetables except for the spinach, and stir-fry for 3 to 5 minutes, or until tender-crisp. At the last minute, add the spinach and cook until wilted. Add the vinegar and season with salt and pepper to taste. Serve over the quinoa.

· Lentil and Vegetable Soup ·

Makes 4 servings

The beauty of lentils is that they are high in protein, like beans, but they cook much quicker. This soup also features spinach and carrots, which make it a high-GSH meal. Serve with Field of Greens Salad (page 220), if desired.

1 cup dried green or brown lentils, rinsed

½ green bell pepper, seeded and chopped

½ red bell pepper, seeded and chopped

½ medium-size onion, diced

2 celery stalks, diced

1 carrot, peeled and diced

2 cups low-sodium vegetable or chicken stock

1 garlic clove, finely chopped

½ teaspoon finely chopped fresh chives

Sea salt and freshly ground black pepper

3 cups baby or chopped large-leaf spinach

Place all ingredients, except the salt, pepper, and spinach, in a large soup pot. Add 6 cups of water. Bring to a boil, then lower heat and let simmer until the lentils are tender, 15 to 20 minutes. Add water as needed (the broth should be light, not thick). Adjust the salt and pepper to taste and add the spinach. Stir until the spinach wilts and serve hot.

· Pan-Seared Fish with Raspberry Vinaigrette ·

Makes 2 servings

The vinaigrette served with this dish allows you to sneak some vitamin C–rich raspberries into your meal. When selecting your fish, keep in mind which are lowest in mercury (see page 148). Serve with a side of Field of Greens Salad (page 220) and, if desired, a side of quinoa or baked potato.

2 fish fillets

Sea salt and freshly ground black pepper

2 tablespoons extra-virgin olive oil

2 rosemary sprigs

1 garlic clove, chopped

Raspberry Vinaigrette (recipe follows)

Salt and pepper the fillets as desired. In a medium-size sauté pan, heat the olive oil over medium heat. Add the rosemary and garlic to the oil and cook for 1 minute, being careful not to let the garlic brown. Remove the rosemary sprigs and add the fish to the pan. Sear the fillets on both sides until the fish flakes easily. (The time will vary depending on the thickness of the fish). Place the fish on a serving platter and drizzle the raspberry vinaigrette on top.

Raspberry Vinaigrette

½ cup fresh raspberries

1½ tablespoons red wine vinegar

2 teaspoons stevia

½ teaspoon Dijon mustard

¼ cup extra-virgin olive oil

Place all the ingredients in a blender and blend thoroughly.

ACKNOWLEDGMENTS

I COULD THANK QUITE AN ENDLESS LIST OF PEOPLE WHO HAVE HAD AN IM-pact in my life. But to isolate those who have most directly been a part of this book, I'll do my best to list them all here.

To my father, you have been my moral compass and my example of a selfless, generous spirit. Thank you for teaching me that I am on this planet to give and serve, and that all of my daily needs will be taken care of if I remember that. I respect you more than words can convey.

I also want to honor my late mother, who inspired me to be selfless and to follow in her footsteps of having a service mindset.

Thank you to my wife, Priya, and our three children, Akhil, Amani, and Sahana. Thank you for making me laugh on a daily basis! You are my life-lines and the people that I look forward to seeing the most at the end of each day. You make it all worth it.

I'd also like to thank my two brothers, who have always been there for me to offer a hand and pick me up through any challenging times.

Dr. Chinh Tran, my fellow USC alumnus, who over the last twenty years has become not only a trusted lab partner and cocreator, but also like a surrogate family member: Without you, I would not be where I am.

Sitting down to write a medically based book rooted in molecular biology seemed like a very daunting prospect to me. Until I met Daryn Eller. Daryn helped me strategize, structure, and write the words on these pages. She captured not just the tone of my voice but my thought process and the feelings I had about all the topics we covered. She is a consummate and diligent fact-checker, and writes with heart and integrity. Most of all, Daryn is an authentic human. She made this writing process possible, and I would not have wanted anyone else on this journey with me. I thank you, Daryn, for putting my thoughts, findings, and research into words that everyone, not just the medical community, can understand and embrace.

From the bottom of my heart, to all the doctors, fellow pharmacists, professors, and colleagues along the way that have contributed knowledge, expertise, time, and/or passion, I am so grateful that we share the like-minded mission of offering pathways of healing to our global society.

To my indirect and direct family members who have supported me and been a part of this journey. You know who you are, and I humbly thank every one of you.

A special thank-you to Michele Roland, who has been one of my biggest supporters and a passionate fan of glutathione. She prodded me to write this book for quite some time, and jumped in to help strategize and research sections—often pulling all-nighters proofreading and fact-checking. And thank you for introducing me to Lindi Stoler, who introduced me to my wonderful literary agent, Steve Troha.

Thank you, Steve and Jan Baumer from Folio Literary Agency. You made this a very painless process.

Also, a big thanks to Renee Sedliar and her editing team at Hachette—your belief, diligence, and encouragement have been incredibly refreshing.

Last, and probably most important, I'd like to thank you, the reader, for being your own health advocate and willing to learn some new things that will most likely improve your health and overall quality of life. I realize that time tends to be our biggest commodity, so I appreciate you using your precious time to soak in the information on these pages. And I hope that what you learn will allow you to be a conduit through which important information can flow to others as well. I believe that if we tackle our health one by one, then share the knowledge, our families, communities, and maybe even the entire world will become a better, healthier place.

SOURCES

INTRODUCTION

Andersen, H. R., B. Jeune, H. Nybo, J. B. Nielsen, K. Andersen-Ranberg, and P. Grandjean. "Low Activity of Superoxide Dismutase and High Activity of Glutathione Reductase in Erythrocytes from Centenarians." *Age and Ageing* 27, no. 5 (1998): 643–648.

Erden-Inal, Mine, Emine Sunal, and Güngör Kanbak. "Age-Related Changes in the Glutathione Redox System." *Cell Biochemistry and Function* 20, no. 1 (March 2002): 61–66.

LaValle, James. Interview with author, February 7, 2020.

CHAPTER 1

"Antioxidants: More Is Not Always Better." *Consumer Reports*, March 2013. https://www.consumerreports.org/cro/2013/03/antioxidants-more-is-not-always-better/index.htm.

Gutman, Jimmy. *GSH: Your Body's Most Powerful Protector*. 3rd ed. Montreal: Kudo.Ca Communications, 2002, 54.

Halliwell, B. "Biochemistry of Oxidative Stress." *Biochemical Society Transactions* 35 (2007): 1147–1150.

Lobo, V., A. Patil, A. Phatak, and N. Chandra. "Free Radicals, Antioxidants and Functional Foods: Impact on Human Health." *Pharmacognosy Reviews* 4, no. 8 (July 1, 2010): 118–126.

"Moderate Levels of 'Free Radicals' Found Beneficial to Healing Wounds." UC San Diego News Center, October 13, 2014. https://ucsdnews.ucsd.edu /pressrelease/moderate_levels_of_free_radicals_found_beneficial_to_healing _wounds.

"Molecule of the Week Archive: Glutathione, May 29, 2017." American Chemical Society. Accessed January 16, 2020. https://www.acs.org/content/acs /en/molecule-of-the- week/archive/g/glutathione.html.

"The Nutrition Source: Vitamin A." Harvard School of Public Health. Accessed January 16, 2020. https://www.hsph.harvard.edu/nutritionsource /vitamin-a/.

Packer, Lester, and Carol Colman. *The Antioxidant Miracle.* New York: John Wiley & Sons, 1999, 23, 31.

Simoni, R. D., R. L. Hill, and M. Vaughn. "The Discovery of Glutathione by F. Gowland Hopkins and the Beginning of Biochemistry at Cambridge University." *Journal of Biological Chemistry* 277, no. 24 (June 14, 2002): e13.

"Sir Frederick Hopkins, Biographical." The Nobel Prize. Accessed January 16, 2020. https://www.nobelprize.org/prizes/medicine/1929/hopkins/biographical/.

"The Top 6 Antioxidant Creams and Serums." *Allure*, June 25, 2012. https:// www.allure.com/gallery/top-6-antioxidant-creams-and-serums.

CHAPTER 2

Amie. Email message to author, November 16, 2019.

"Chronic Inflammation." National Center for Biotechnology Information, last updated December 13, 2019. https://www.ncbi.nlm.nih.gov/books /NBK493173/.

Erden-Inal, Mine, Emine Sunal, and Güngör Kanbak. "Age-Related Changes in the Glutathione Redox System." *Cell Biochemistry and Function* 20, no. 1 (March 2002): 61–66.

Ferguson, Gavin, and Wallace Bridge. "Glutamate Cysteine Ligase and the Age-Related Decline in Cellular Glutathione: The Therapeutic Potential of γ-Glutamylcysteine." *Archives of Biochemistry and Biophysics* 593 (March 1, 2016): 12–23.

Freed, Rachel, Cecilia Hollenhorst, Xiangling Mao, Nora Weiduschat, Dikoma Shungu, and Vilma Gabbay. "946. Decreased Occipital Glutathione in

Adolescent Depression: A Magnetic Resonance Spectroscopy Study." *Biological Psychiatry* 81, no. 10 (May 15, 2017): S383–S383.

Frye, Richard E., Stepan Melnyk, George Fuchs, Tyra Reid, Stefanie Jernigan, Oleksandra Pavliv, Amanda Hubanks, David W. Gaylor, Laura Walters, and S. Jill James. "Effectiveness of Methylcobalamin and Folinic Acid Treatment on Adaptive Behavior in Children with Autistic Disorder Is Related to Glutathione Redox Status." *Autism Research and Treatment* 2013 (2013).

"Genetics Home Reference: Glutathione Synthetase Deficiency." NIH, US National Library of Medicine, last updated January 7, 2020. https://ghr.nlm.nih.gov/condition/glutathione-synthetase-deficiency#synonyms.

Ghezzi, Pietro, and P. Ghezzi. "Role of Glutathione in Immunity and Inflammation in the Lung." *International Journal of General Medicine* 4 (January 2011): 105–113.

Gibson, Sara A., Željka Korade, and Richard C. Shelton. "Oxidative Stress and Glutathione Response in Tissue Cultures from Persons with Major Depression." *Journal of Psychiatric Research* 46, no. 10 (October 2012): 1326–1332.

"Glutathione Blood Test." Discounted Labs. Accessed January 18, 2020. https://www.discountedlabs.com/glutathione-blood-test.

Guthrie, Catherine. "Glutathione: The Great Protector." Experience L!fe, May 2015. https://experiencelife.com/article/glutathione-the-great-protector-2/.

"Inflammation: A Unifying Theory of Disease?" Harvard Health Letter, April 2006. https://www.health.harvard.edu/newsletter_article/Inflammation_A_unifying_theory_of_disease.

James, S. Jill, Stepan Melnyk, Stefanie Jernigan, Mario A. Cleves, Charles H. Halsted, Donna H. Wong, Paul Cutler, et al. "Metabolic Endophenotype and Related Genotypes Are Associated with Oxidative Stress in Children with Autism." *American Journal of Medical Genetics Part B: Neuropsychiatric Genetics* 141B, no. 8 (December 5, 2006): 947–956.

Miller, Molly. "Toxic Exposure: Chemicals Are in Our Water, Food, Air and Furniture." University of California San Francisco, June 22, 2017. https://www.ucsf.edu/news/2017/06/407416/toxic-exposure-chemicals-are-our-water-food-air-and-furniture.

"A New Approach to Parkinson's?" Andrew Weil, MD. March 10, 2009. https://www.drweil.com/health-wellness/body-mind-spirit/disease-disorders/a-new-approach-to-parkinsons/.

"New CDC Report: More Than 100 Million Americans Have Diabetes or Prediabetes." Centers for Disease Control and Prevention, July 18, 2017. https://www.cdc.gov/media/releases/2017/p0718-diabetes-report.html.

"Researchers Unravel Role of Oxidative Stress in Autism Spectrum Disorder." Children's Hospital Los Angeles, July 3, 2013. https://www.chla.org/press-release/researchers-unravel-role-oxidative-stress-autism-spectrum-disorder.

Rose, S., S. Melnyk, O. Pavliv, S. Bai, T. G. Nick, R. E. Frye, and S. J. James. "Evidence of Oxidative Damage and Inflammation Associated with Low Glutathione Redox Status in the Autism Brain." *Translational Psychiatry* 2, no. 7 (July 10, 2012): e134.

Salim, Samina, Mohammad Asghar, Manish Taneja, Iiris Hovatta, Gaurav Chugh, Craig Vollert, and Anthony Vu. "Potential Contribution of Oxidative Stress and Inflammation to Anxiety and Hypertension." *Brain Research* 1404 (2011): 63–71.

Sekhar, Rajagopal V., Siripoom V. Mckay, Sanjeet G, Patel, Anuradha P. Guthikonda, Vasumathi T. Reddy, Ashok Balasubramanyam, and Jahoor Farook. "Glutathione Synthesis Is Diminished in Patients with Uncontrolled Diabetes and Restored by Dietary Supplementation with Cysteine and Glycine." *Diabetes Care* 34, no. 1 (January 1, 2011): 162–167.

Shimizu, Haruki, Yutaka Kiyohara, Isao Kato, Takanari Kitazono, Yumihiro Tanizaki, Michiaki Kubo, Hirofumi Ueno, Setsuro Ibayashi, Masatoshi Fujishima, and Mitsuo Iida. "Relationship Between Plasma Glutathione Levels and Cardiovascular Disease in a Defined Population: The Hisayama Study." *Stroke* 35, no. 9 (September 2004): 2072–2077.

Tapiero, H., D. M. Townsend, and K. D. Tew. "The Antioxidant Role of Selenium and Seleno-Compounds." *Biomedicine & Pharmacotherapy* 57, no. 3–4 (2003): 134–144.

"Understanding Acetaminophen Poisoning." Science Daily, October 14, 2002. https://www.sciencedaily.com/releases/2002/10/021014072451.htm.

"UV Exposure Has Increased over the Last 30 Years, but Stabilized Since the Mid-1990s." NASA, March 3, 2010. https://www.nasa.gov/topics/solarsystem/features/uv-exposure.html.

Vogt, Barbara L., and John P. Richie. "Glutathione Depletion and Recovery After Acute Ethanol Administration in the Aging Mouse." *Biochemical Pharmacology* 73, no. 10 (2007): 1613–1621.

Wang, Lulu W., Daniel J. Tancredi, and Dan W. Thomas. "The Prevalence of Gastrointestinal Problems in Children Across the United States with Autism Spectrum Disorders from Families with Multiple Affected Members." *Journal of Developmental & Behavioral Pediatrics* 32, no. 5 (June 2011): 351–360.

"What Is an Inflammation?" National Center for Biotechnology Information, last updated February 22, 2018. https://www.ncbi.nlm.nih.gov/books /NBK279298/.

CHAPTER 3

"2019–2020 U.S. Flu Season: Preliminary Burden Estimates." Centers for Disease Control and Prevention, last updated February 14, 2020. https://www .cdc.gov/flu/about/burden/preliminary-in-season-estimates.htm.

Adams, James D., Lori K. Klaidman, Ifeoma N. Odunze, Howard C. Shen, and Carol A. Miller. "Alzheimer's and Parkinson's Disease. Brain Levels of Glutathione, Glutathione Disulfide, and Vitamin E." *Molecular and Chemical Neuropathology* 14 (June 1991): 213–226.

"Alzheimer's and Dementia Facts and Figures." Alzheimer's Association. Accessed January 21, 2020. https://www.alz.org/alzheimers-dementia/facts-figures.

"Antioxidant Eye Drops Provide Another Option for Cataract Patients." Healio, October 2015. https://www.healio.com/optometry/cataract-surgery/news /print/primary-care-optometry-news/%7B1388d46d-8c21-4da3-9c27-b02c180 ef341%7D/antioxidant-eye-drops-provide-another-option-for-cataract-patients.

Aoyama, Koji, and Toshio Nakaki. "Impaired Glutathione Synthesis in Neurodegeneration." *International Journal of Molecular Sciences* 14, no. 10 (October 1, 2013): 21021–21044.

Ashfaq, Salman, Jerome L. Abramson, Dean P. Jones, Steven D. Rhodes, William S. Weintraub, W. Craig Hooper, Vaccarino Viola, David G. Harrison, and Arshed A. Quyyumi. "The Relationship Between Plasma Levels of Oxidized and Reduced Thiols and Early Atherosclerosis in Healthy Adults." *Journal of the American College of Cardiology* 47, no. 5 (March 7, 2006): 1005–1011.

Ayala, Antonio, Mario F. Muñoz, and Sandro Argüelles. "Lipid Peroxidation: Production, Metabolism, and Signaling Mechanisms of Malondialdehyde and 4-Hydroxy-2-Nonenal." *Oxidative Medicine and Cellular Longevity* 2014, no. 4 (January 1, 2014): 360438.

Babu, G. Nagesh, Alok Kumar, Ramesh Chandra, S. K. Puri, R. L. Singh, Kalita Jayantee, and U. K. Misra. "Oxidant–Antioxidant Imbalance in the Erythrocytes of Sporadic Amyotrophic Lateral Sclerosis Patients Correlates with the Progression of Disease." *Neurochemistry International* 52, no. 6 (2008): 1284–1289.

Blum, Susan. *The Immune System Recovery Plan.* New York: Scribner, 2013, 250.

Cai, Jiyang, Yan Chen, Shaguna Seth, Furukawa Satoru, Richard W. Compans, and Dean P. Jones. "Inhibition of Influenza Infection by Glutathione." *Free Radical Biology and Medicine* 34, no. 7 (April 1, 2003): 928–936.

"Cancer Statistics." National Cancer Institute, updated February 9, 2015. https://www.cancer.gov/about-cancer/understanding/statistics.

Cao, Ruoqiong, Garrett Teskey, Hicret Islamoglu, Rachel Abrahem, Shalok Munjal, Karo Gyurjian, Li Zhong, and Vishwanath Venketaraman. "Characterizing the Effects of Glutathione as an Immunoadjuvant in the Treatment of Tuberculosis." *Antimicrobial Agents and Chemotherapy* 62, no. 11 (November 1, 2018).

Damy, Thibaud, Matthias Kirsch, Lara Khouzami, Philippe Caramelle, Philippe Le Corvoisier, Françoise Roudot-Thoraval, Jean-Luc Dubois-Randé, et al. "Glutathione Deficiency in Cardiac Patients Is Related to the Functional Status and Structural Cardiac Abnormalities (Glutathione in Heart Diseases)." *PLoS ONE* 4, no. 3 (March 25, 2009): e4871.

De Flora, S., C. Grassi, and L. Carati. "Attenuation of Influenza-Like Symptomatology and Improvement of Cell-Mediated Immunity with Long-Term N-Acetylcysteine Treatment." *European Respiratory Journal* 10, no. 7 (July 1, 1997): 1535–1541.

"Disease." Lexico. Accessed January 21, 2020. https://www.lexico.com/en/definition/disease.

Dong, Justin, Anand Muthiah, Parveen Hussain, Miya Yoshida, and Vishwanath Venketaraman. "Selenium Supplementation, Antioxidant Effects, and Immune Restorative Effects in Human Immunodeficiency Virus." In *HIV/AIDS: Oxidative Stress and Dietary Antioxidants*, 197–205. Edited by Victor R. Preedy and Ronald Ross Wilson. London: Academic Press, 2018.

Dringen, Ralf. "Metabolism and Functions of Glutathione in Brain." *Progress in Neurobiology* 62, no. 6 (2000): 649–671.

Ferlita, Steve, Aram Yegiazaryan, Navid Noori, Gagandeep Lal, Nguyen Timothy, Kimberly To, and Vishwanath Venketaraman. "Type 2 Diabetes Mellitus and Altered Immune System Leading to Susceptibility to Pathogens, Especially *Mycobacterium tuberculosis*." *Journal of Clinical Medicine* 8, no. 12 (December 16, 2019).

"Frequently Asked Questions: Cancer Facts." Cancer Research Institute. Accessed January 22, 2020. https://www.cancerresearch.org/faq.

G., Kenneth. Interview with author, November 4, 2019.

Gutman, Jimmy. *GSH: Your Body's Most Powerful Protector.* 3rd ed. Montreal: Kudo.Ca Communications, 2002, 43.

"Hashimoto's Disease." Mayo Clinic. Accessed January 21, 2020. https://www.mayoclinic.org/diseases-conditions/hashimotos-disease/symptoms-causes/syc-20351855.

Hauser, Robert A., Kelly E. Lyons, Terry McClain, Summer Carter, and David Perlmutter. "Randomized, Double-Blind, Pilot Evaluation of Intravenous Glutathione in Parkinson's Disease." *Movement Disorders* 24, no. 7 (May 15, 2009): 979–983.

Herzenberg, Leonore A., Stephen C. De Rosa, J. Gregson Dubs, Mario Roederer, Michael T. Anderson, Stephen W. Ela, Stanley C. Deresinski, and Leonard A. Herzenberg. "Glutathione Deficiency Is Associated with Impaired Survival in HIV Disease." *Proceedings of the National Academy of Sciences of the United States of America* 94, no. 5 (March 4, 1997): 1967–1972.

Ho, Min-Chieh, Yi-Jie Peng, Shih-Jen Chen, and Shih-Hwa Chiou. "Senile Cataracts and Oxidative Stress." *Journal of Clinical Gerontology and Geriatrics* 1, no. 1 (September 2010): 17–21.

Honda, Yasushi, Takaomi Kessoku, Yoshio Sumida, Takashi Kobayashi, Takayuki Kato, Yuji Ogawa, Wataru Tomeno, et al. "Efficacy of Glutathione for the Treatment of Nonalcoholic Fatty Liver Disease: An Open-Label, Single-Arm, Multicenter, Pilot Study. (Report)." *BMC Gastroenterology* 17, no. 1 (August 8, 2017): 96.

"The Hong Kong Foot." Feet for Health, November 16, 2014. http://feetforhealth.blogspot.com/2014/11/the-hong-kong-foot.html.

Hou, Yujun, Xiuli Dan, Mansi Babbar, Yong Wei, Steen G. Hasselbalch, Deborah L. Croteau, and Vilhelm A. Bohr. "Ageing as a Risk Factor for Neurodegenerative Disease." *Nature Reviews. Neurology* 15, no. 10 (October 1, 2019): 565–581.

Hyman, Mark. "Glutathione: The Mother of All Antioxidants." Huffington Post, last updated November 17, 2011. https://www.huffpost.com/entry/glutathione-the-mother-of_b_530494.

Kathy. Email message to author, November 4, 2019.

Lund, Troy, and Dan Purser. *Real Glutathione* (n.p., 2014), chap. 2, 334. Kindle.

Mackay, Ian R., Fred S. Rosen, Anne Davidson, and Betty Diamond. "Autoimmune Diseases." *New England Journal of Medicine* 345, no. 5 (August 2, 2001): 340–350.

Mandal, Pravat K., Manjari Tripathi, and Sreedevi Sugunan. "Brain Oxidative Stress: Detection and Mapping of Anti-Oxidant Marker 'Glutathione' in Different Brain Regions of Healthy Male/Female, MCI and Alzheimer Patients

Using Non-Invasive Magnetic Resonance Spectroscopy." *Biochemical and Biophysical Research Communications* 417, no. 1 (January 6, 2012): 43–48.

"Medline Plus: Fatty Liver Disease." US National Library of Medicine. Accessed January 22, 2020. https://medlineplus.gov/fattyliverdisease.html.

Minette Lagman, Judy Ly, Tommy Saing, Manpreet Kaur Singh, Enrique Vera Tudela, Devin Morris, Po-Ting Chi, Cesar Ochoa, Airani Sathananthan, and Vishwanath Venketaraman. "Investigating the Causes for Decreased Levels of Glutathione in Individuals with Type II Diabetes." *PLoS ONE* 10, no. 3 (January 1, 2015): e0118436.

Morris, Devin, Carlos Guerra, Clare Donohue, Hyoung Oh, Melissa Khurasany, and Vishwanath Venketaraman. "Unveiling the Mechanisms for Decreased Glutathione in Individuals with HIV Infection." *Clinical & Developmental Immunology* 2012 (2012): 734125–734125.

Morris, Devin, Melissa Khurasany, Nguyen Thien, John Kim, Frederick Guilford, Rucha Mehta, Dennis Gray, Beatrice Saviola, and Vishwanath Venketaraman. "Glutathione and Infection." *BBA—General Subjects* 1830, no. 5 (May 2013): 3329–3349.

"Neurodegenerative Diseases." National Institute of Environmental Health Sciences. Accessed January 21, 2020. https://www.niehs.nih.gov/research/supported/health/neurodegenerative/index.cfm.

"New CDC Report: More Than 100 Million Americans Have Diabetes or Prediabetes." Centers for Disease Control and Prevention, July 18, 2017. https://www.cdc.gov/media/releases/2017/p0718-diabetes-report.html.

Nguyen, Dan, Susan L. Samson, Vasumathi T. Reddy, Erica V. Gonzalez, and Rajagopal V. Sekhar. "Impaired Mitochondrial Fatty Acid Oxidation and Insulin Resistance in Aging: Novel Protective Role of Glutathione." *Aging Cell* 12, no. 3 (June 2013): 415–425.

Packer, Lester, and Carol Colman. *The Antioxidant Miracle.* New York: John Wiley & Sons, 1999, 21.

Perricone, Carlo, Caterina De Carolis, and Roberto Perricone. "Glutathione: A Key Player in Autoimmunity." *Autoimmunity Reviews* 8, no. 8 (2009): 697–701.

Pizzino, Gabriele, Natasha Irrera, Mariapaola Cucinotta, Giovanni Pallio, Federica Mannino, Vincenzo Arcoraci, Francesco Squadrito, Domenica Altavilla, and Alessandra Bitto. "Oxidative Stress: Harms and Benefits for Human Health." *Oxidative Medicine and Cellular Longevity* 2017 (2017): 13.

"Putting a Stop to Leaky Gut." Harvard Health Publishing, December 2018. https://www.health.harvard.edu/diseases-and-conditions/putting-a-stop-to-leaky-gut.

Qian, Li, Wenjun Wang, Yan Zhou, and Jianxin Ma. "Effects of Reduced Glutathione Therapy on Chronic Hepatitis B." *Central-European Journal of Immunology* 42, no. 1 (2017): 97–100.

Rae, Caroline D., and Stephen R. Williams. "Glutathione in the Human Brain: Review of Its Roles and Measurement by Magnetic Resonance Spectroscopy." *Analytical Biochemistry* 529 (July 15, 2017): 127–143.

Riederer, Peter, Emin Sofic, Wolf-Dieter Rausch, Bruno Schmidt, Gavin P. Reynolds, Kurt Jellinger, and Moussa B. H. Youdim. "Transition Metals, Ferritin, Glutathione, and Ascorbic Acid in Parkinsonian Brains." *Journal of Neurochemistry* 52, no. 2 (February 1989): 515–520.

Rosangela, F. F. de Araujo, Danyelly Bruneska, G. Martins, Maria Amelia C. S. M. Borba. "Oxidative Stress and Disease." In *A Master Regulator of Oxidative Stress—The Transcription Factor Nrf2*, edited by Eduardo Osiris Madrigal-Santillan, Angel Morales-Gonzalez, and Jose Antonio Morales-Gonzalez. InTechOpen, 2016. https://www.intechopen.com/books/a-master-regulator-of-oxidative-stress-the-transcription-factor-nrf2/oxidative-stress-and-disease.

Rostami, R., M. R. Aghasi, A. Mohammadi, and J. Nourooz-Zadeh. "Enhanced Oxidative Stress in Hashimoto's Thyroiditis: Inter-Relationships to Biomarkers of Thyroid Function." *Clinical Biochemistry* 46, nos. 4–5 (March 2013): 308–312.

Samiec, Paula S., Carolyn Drews-Botsch, Elaine W. Flagg, Joanne C. Kurtz, Paul Sternberg, Robyn L. Reed, and Dean P. Jones. "Glutathione in Human Plasma: Decline in Association with Aging, Age-Related Macular Degeneration, and Diabetes." *Free Radical Biology and Medicine* 24, no. 5 (1998): 699–704.

Schimel, Andrew M., Linu Abraham, Douglas Cox, Abdoulaye Sene, Courtney Kraus, Dru S. Dace, Nuran Ercal, and Rajendra S. Apte. "N-Acetylcysteine Amide (NACA) Prevents Retinal Degeneration by Up-Regulating Reduced Glutathione Production and Reversing Lipid Peroxidation." *American Journal of Pathology* 178, no. 5 (2011): 2032–2043.

Sekhar, Rajagopal V., Siripoom V. Mckay, Sanjeet G. Patel, Anuradha P. Guthikonda, Vasumathi T. Reddy, Ashok Balasubramanyam, and Farook Jahoor. "Glutathione Synthesis Is Diminished in Patients with Uncontrolled Diabetes and Restored by Dietary Supplementation with Cysteine and Glycine." *Diabetes Care* 34, no. 1 (January 2011): 162–167.

Shimizu, Haruki, Yutaka Kiyohara, Isao Kato, Takanari Kitazono, Yumihiro Tanizaki, Michiaki Kubo, Hirofumi Ueno, Setsuro Ibayashi, Masatoshi Fujishima, and Mitsuo Iida. "Relationship Between Plasma Glutathione Levels and Cardiovascular Disease in a Defined Population: The Hisayama Study." *Stroke* 35, no. 9 (September 2004): 2072–2077.

Smeyne, Michelle, and Richard Jay Smeyne. "Glutathione Metabolism and Parkinson's Disease." *Free Radical Biology and Medicine* 62 (September 2013): 13–25.

Solano, Ray. Interview with author, September 12, 2019.

TEDMED. "Mark Hyman at TEDMED 2010." YouTube video, 19:09. Posted December 7, 2010. https://www.youtube.com/watch?v=IhkLcpJTV9M &app=desktop.

"The Top 10 Causes of Death." World Health Organization, May 24, 2018. https://www.who.int/news-room/fact-sheets/detail/the-top-10-causes-of-death.

Tokarz, Paulina, Kai Kaarniranta, and Janusz Blasiak. "Role of Antioxidant Enzymes and Small Molecular Weight Antioxidants in the Pathogenesis of Age-Related Macular Degeneration (AMD)." *Biogerontology* 14, no. 5 (October 2013): 461–482.

Traverso, Nicola, Roberta Ricciarelli, Mariapaola Nitti, Barbara Marengo, Anna Lisa Furfaro, Maria Adelaide Pronzato, Umberto Maria Marinari, and Cinzia Domenicotti. "Role of Glutathione in Cancer Progression and Chemoresistance." *Oxidative Medicine and Cellular Longevity* 2013 (2013): 10.

Venketaraman, Vishwanath. Interview with author, September 5, 2020.

"What Is Cancer?" National Cancer Institute, updated February 9, 2015. https://www.cancer.gov/about-cancer/understanding/what-is-cancer.

"What Is Viral Hepatitis?" Centers for Disease Control and Prevention, last updated October 30, 2019. https://www.cdc.gov/hepatitis/abc/index.htm.

"What We Eat Makes a Big Difference." Letter to the Editor. *New York Times*, September 9, 2019.

World Health Organization. *Global Tuberculosis Report: Executive Summary 2018*. Geneva: World Health Organization, 2018, 1.

CHAPTER 4

"10 People and Chemicals in Their Midst." Environmental Defense Fund, May 2017. https://www.edf.org/health/10-people-and-chemicals-their-midst.

"Alcohol Mechanism: An Update." National Institute on Alcohol Abuse and Alcoholism, July 2007. https://pubs.niaaa.nih.gov/publications/aa72/aa72.htm.

Bryanna. Interview with author, September 25, 2019.

CEO. Email message to author, December 5, 2019.

"Dioxins." National Institute of Environmental Health Sciences. Accessed February 8, 2020. https://www.niehs.nih.gov/health/topics/agents/dioxins/index .cfm.

"Fact Sheet: Cancer Risk in Firefighting." National Fire Protection Association, February 3, 2017. https://www.nfpa.org/-/media/Files/Code-or-topic-fact -sheets/FactSheetFFLungCancer.pdf.

"Findings from a Study of Cancer Among U.S. Fire Fighters." Centers for Disease Control and Prevention, July 2016. https://www.cdc.gov/niosh/pgms /worknotify/pdfs/ff-cancer-factsheet-final-508.pdf.

"Firefighter Cancer Fact Check." Firefighter Close Calls. Accessed February 8, 2020. https://www.firefighterclosecalls.com/wp-content/uploads/2017/06/FF -Cancer-Fact-Sheet.pdf.

"Flame Retardants." National Institute of Environmental Health Science. Accessed January 24, 2020. https://www.niehs.nih.gov/health/materials /flame_retardants_508.pdf.

"Formaldehyde." American Cancer Society, last updated May 23, 2014. https://www.cancer.org/cancer/cancer-causes/formaldehyde.html.

Garling, Kevin. Email message to author, April 20, 2020.

Godwill, Azeh Engwa, Paschaline Udoka Ferdinand, Friday Nweke Nwalo, and Marian N. Unachukwu. "Mechanism and Health Effects of Heavy Metal Toxicity." In *Poisoning in the Modern World—New Tricks for an Old Dog?* Edited by Ozgur Karcioglu. IntechOpen, 2019.

Grubben, M. J. A. L., C. C. M. Den Braak, F. M. Nagengast, and W. H. M. Peters. "Low Colonic Glutathione Detoxification Capacity in Patients at Risk for Colon Cancer." *European Journal of Clinical Investigation* 36, no. 3 (March 2006): 188–192.

Gutman, Jimmy. *GSH: Your Body's Most Powerful Protector.* 3rd ed. Montreal: Kudo.Ca Communications, 2002, 26.

Hirsch, Jesse. "Arsenic and Lead Are in Your Fruit Juice: What You Need to Know." *Consumer Reports*, last updated January 30, 2019. https://www .consumerreports.org/food-safety/arsenic-and-lead-are-in-your-fruit-juice-what-you -need-to-know/#chart.

"How Does the Liver Work?" National Center for Biotechnology Information, last updated August 22, 2016. https://www.ncbi.nlm.nih.gov/books/NBK279393/.

"Key Findings: The State of the Air 2019." American Lung Association. Accessed February 7, 2020. https://www.lung.org/our-initiatives/healthy-air/sota /key-findings/.

Mercola, Joseph. "Farmed Salmon = the Most Toxic Food in the World." Organic Consumers, July 25, 2018. https://www.organicconsumers.org/news/farmed-salmon-toxic-flame-retardants.

Minnick, Fred. "In Search of a Cure for the Dreaded Hangover." *Scientific American*, March 17, 2014. https://www.scientificamerican.com/article/in-search-of-a-cure-for-the-dreaded-hangover/.

"More Americans Take Prescription Medication." Medscape. Accessed February 7, 2020. https://www.medscape.com/viewarticle/500164.

Olsen, Erik D. "The Broken Safe Drinking Water Act Won't Fix the PFAS Crisis." NRDC, September 12, 2019. https://www.nrdc.org/experts/erik-d-olson/broken-safe-drinking-water-act-wont-fix-pfas-crisis.

Ozguner, Fehmi, Ahmet Altinbas, Mehmet Ozaydin, Abdullah Dogan, Huseyin Vural, A. Nesimi Kisioglu, Gokhan Cesur, and Nurhan Gumral Yildirim. "Mobile Phone-Induced Myocardial Oxidative Stress: Protection by a Novel Antioxidant Agent Caffeic Acid Phenethyl Ester." *Toxicology and Industrial Health* 21, no. 9 (August 2005): 223–230.

"PCBs in Fish and Shellfish." Environmental Defense Fund. Accessed February 7, 2020. http://seafood.edf.org/pcbs-fish-and-shellfish.

"Per- and Polyfluoroalkyl Substances (PFAS) and Your Health." Agency for Toxic Substances and Disease Registry, last updated January 21, 2020. https://www.atsdr.cdc.gov/pfas/health-effects.html.

"Perchloroethylene (PCE, PERC)." Tox Town. US Library of Medicine. Accessed January 25, 2020. https://toxtown.nlm.nih.gov/chemicals-and-contaminants/perchloroethylene-pce-perc.

"Perfluoroalkyl and Polyfluoroalkyl Substances (PFAS)." National Institute of Environmental Health Sciences. Accessed January 24, 2020. https://www.niehs.nih.gov/health/topics/agents/pfc/index.cfm.

Perry, Philip. "You'll Never Guess How Many Chemicals Are Inside Your Body Right Now." Big Think, August 30, 2017. https://bigthink.com/philip-perry/youll-never-guess-how-many-chemicals-are-inside-your-body-right-now.

Pinkus, Lawrence M., Jeanne N. Ketley, and William B. Jakoby. "The Glutathione S-Transferases as a Possible Detoxification System of Rat Intestinal Epithelium." *Biochemical Pharmacology* 26, no. 24 (1977): 2359–2363.

Pizzorno, Joseph. "Glutathione!" *Integrative Medicine* (Encinitas, CA) 13, no. 1 (February 2014): 8–12.

Purser, Dan, and R. Tillotson. "34 Volunteer Nanoized GSH Double Blinded Detox Study." Shared with author, September 2019.

"Residential Furniture Survey." Center for Environmental Health. Accessed January 24, 2020. https://www.ceh.org/residential-furniture/.

Swift, Robert, and Dena Davidson. "Alcohol Hangover: Mechanisms and Mediators." *Alcohol Health and Research World* 22, no. 1 (January 1, 1998): 54–60.

Tchounwou, P. B., C. G. Yedjou, A. K. Patlolla, and D. J. Sutton. "Heavy Metal Toxicity and the Environment." *Experientia Supplementum* (2012): 101, 133–164.

"Three Ways 'BPA-Free' Won't Protect You." Environmental Defense Fund. Accessed January 24, 2020. https://www.edf.org/health/three-ways-bpa-free-wont-protect-you.

"Toxic Chemicals." Natural Resources Defense Council. Accessed February 7, 2020. https://www.nrdc.org/issues/toxic-chemicals.

"Toxic Substances Portal—Formaldehyde." Agency for Toxic Substances & Disease Registry, last updated September 2008. https://www.atsdr.cdc.gov/phs/phs.asp?id=218&tid=39.

"What Are Parabens, and Why Don't They Belong in Cosmetics?" Environmental Working Group, April 9, 2019. https://www.ewg.org/californiacosmetics/parabens.

"What Does the Liver Do?" Medical News Today, last updated March 2, 2018. https://www.medicalnewstoday.com/articles/305075.php.

"What Is Azodicarbonamide? 9 Exposing Facts." Global Healing Center, last updated June 18, 2014. https://www.globalhealingcenter.com/natural-health/what-is-azodicarbonamide/.

Zero Breast Cancer. "Phthalates: The Everywhere Chemical." National Institute of Environmental Health Science. Accessed January 24, 2020. https://www.niehs.nih.gov/research/supported/assets/docs/j_q/phthalates_the_everywhere._chemical_handout_508.pdf.

CHAPTER 5

Bartlett, Zane. "The Hayflick Limit." The Embryo Project Encyclopedia, November 14, 2014. https://embryo.asu.edu/pages/hayflick-limit.

"Blackburn Lab Research." University of California, San Francisco. Accessed February 12, 2020. https://blackburnlab.ucsf.edu/.

Borrás, Consuelo, Juan M. Esteve, Juan R. Viña, Juan Sastre, José Viña, and Federico V. Pallardó. "Glutathione Regulates Telomerase Activity in 3T3

Fibroblasts." *Journal of Biological Chemistry* 279, no. 33 (August 13, 2004): 34332–34335.

Brack, C., E. Bechter-Thüring, and M. Labuhn. "N-Acetylcysteine Slows Down Ageing and Increases the Life Span of Drosophila melanogaster." *Cellular and Molecular Life Sciences (CMLS)* 53, nos. 11–12 (December 1997): 960–966.

Fernandez, Elizabeth. "Lifestyle Changes May Lengthen Telomeres: A Measure of Cell Aging." University of California, San Francisco, September 16, 2013. https://www.ucsf.edu/news/2013/09/108886/lifestyle-changes-may-lengthen-telomeres-measure-cell-aging.

"Hidden Secret of Immortality Enzyme Telomerase," Science Daily, February 27, 2018. https://www.sciencedaily.com/releases/2018/02/180227142114.htm.

Jin, Kunlin. "Modern Biological Theories of Aging." *Aging and Disease* 1, no. 2 (October 1, 2010): 72–74.

Lang, Calvin A., Betty Jane Mills, Helen L. Lang, Marica C. Liu, Wayne M. Usui, John Richie, Walter Mastropaolo, and Stanley A. Murrell. "High Blood Glutathione Levels Accompany Excellent Physical and Mental Health in Women Ages 60 to 103 Years." *Journal of Laboratory and Clinical Medicine* 140, no. 6 (December 2002): 413–417.

Melinda Wenner Moyer. "The Myth of Antioxidants." *Scientific American,* January 14, 2013: 62–67.

Richie, J. P., B. Mills, and C. A. Lang. "Correction of a Glutathione Deficiency in the Aging Mosquito Increases Its Longevity (42454)." *Proceedings of the Society for Experimental Biology and Medicine* 184, no. 1 (January 1, 1987): 113–117.

"UCSF Profiles: Elizabeth Blackburn, PhD." University of California, San Francisco. Accessed February 12, 2020. https://profiles.ucsf.edu/elizabeth.blackburn.

Yu, Yongjie, Limin Zhou, Yajun Yang, and Yuyu Liu. "Cycloastragenol: An Exciting Novel Candidate for Age-Associated Diseases (Review)." *Experimental and Therapeutic Medicine* 16, no. 3 (September 1, 2018): 2175–2182.

CHAPTER 6

Barranco-Ruiz, Yaira, Jerónimo Aragón-Vela, Cristina Casals, Antonio Martínez-Amat, Emilio Villa-González, and Jesús Huertas. "Chronic Amateur Endurance Practice Improves Oxidative Stress Response for Preserving

Muscle Mass in Senior Adults: 2484 Board #7 June 3, 9:30 AM–11:00 AM." *Medicine & Science in Sports & Exercise* 48, no. 5S, Suppl 1 (May 2016): 683.

Burton, Deborah Anne, Keith Stokes, and George M. Hall. "Physiological Effects of Exercise." *Continuing Education in Anaesthesia Critical Care & Pain* 4, no. 6 (December 2004): 185–188.

Finaud, Julien, Gérard Lac, and Edith Filaire. "Oxidative Stress: Relationship with Exercise and Training." *Sports Medicine* (Auckland, NZ) 36, no. 4 (January 1, 2006): 327–358.

Gutman, Jimmy. *GSH: Your Body's Most Powerful Protector.* 3rd ed. Montreal: Kudo.Ca Communications, 2002, 222.

Hunter, Stephanie. Email message to author, January 31, 2020.

Kerksick, Chad, and Darryn Willoughby. "The Antioxidant Role of Glutathione and N-Acetyl-Cysteine Supplements and Exercise-Induced Oxidative Stress." *Journal of the International Society of Sports Nutrition* 2, no. 2 (December 9, 2005): 38–44.

LaValle, James. Interview with author, February 7, 2020.

Lee, Joohyung, and Priscilla M. Clarkson. "Plasma Creatine Kinase Activity and Glutathione After Eccentric Exercise." *Medicine & Science in Sports & Exercise* 35, no. 6 (June 2003): 930–936.

Medved, I., M. J. Brown, A. R. Bjorksten, K. T. Murphy, A. C. Petersen, S. Sostaric, X. Gong, and M. J. McKenna. "N-Acetylcysteine Enhances Muscle Cysteine and Glutathione Availability and Attenuates Fatigue During Prolonged Exercise in Endurance-Trained Individuals." *Journal of Applied Physiology* 97, no. 4 (October 1, 2004): 1477–1485.

Powers, Scott K., et al. "Exercise-Induced Oxidative Stress: Past, Present and Future." *Journal of Physiology* 594, no. 18 (2016): 5081–5092.

Prousky, Jonathan. "The Treatment of Pulmonary Diseases and Respiratory-Related Conditions with Inhaled (Nebulized or Aerosolized) Glutathione." *Evidence-Based Complementary and Alternative Medicine* 5, no. 1 (March 2008): 27–35.

Puetz, Timothy W., Sara S. Flowers, and Patrick J. O'Connor. "A Randomized Controlled Trial of the Effect of Aerobic Exercise Training on Feelings of Energy and Fatigue in Sedentary Young Adults with Persistent Fatigue." *Psychotherapy and Psychosomatics* 77, no. 3 (March 2008): 167–174.

Reynolds, Gretchen. "Why Vitamins May Be Bad for Your Workout." *New York Times*, February 12, 2014. https://well.blogs.nytimes.com/2014/02/12/why-vitamins-may-be-bad-for-your-workout/?searchResultPosition=2.

Sen, C. K., M. Atalay, and O. Hänninen. "Exercise-Induced Oxidative Stress: Glutathione Supplementation and Deficiency." *Journal of Applied Physiology* 77, no. 5 (November 1994): 2177–2187.

Sofi, Tamara. Interview with author, August 23, 2019.

CHAPTER 7

"Acne." Mayo Clinic. Accessed February 15, 2020. https://www.mayoclinic.org/diseases-conditions/acne/symptoms-causes/syc-20368047.

Basak, Pinar Y., Fatih Gultekin, and Ibrahim Kilinc. "The Role of the Antioxidative Defense System in Papulopustular Acne." *Journal of Dermatology* 28, no. 3 (March 2001): 123–127.

Connor, M. J., and L. A. Wheeler. "Depletion of Cutaneous Glutathione by Ultraviolet Radiation." *Photochemistry and Photobiology* 46, no. 2 (August 1987): 239–45.

El-Domyati, M., S. Attia, F. Saleh, D. Brown, D. E. Birk, F. Gasparro, H. Ahmad, and J. Uitto. "Intrinsic Aging vs. Photoaging: A Comparative Histopathological, Immunohistochemical, and Ultrastructural Study of Skin." *Experimental Dermatology* 11, no. 5 (October 2002): 398–405.

Farage M. A., K. W. Miller, P. Elsner, and H. I. Maibach. "Characteristics of the Aging Skin." *Advanced Wound Care* 2, no. 1 (February 2013): 5–10.

Gordon, Jennifer R. S., and Joaquin C. Brieva. "Unilateral Dermatoheliosis." *New England Journal of Medicine* 366, no. 16 (April 19, 2012): e25.

Gutman, Jimmy. *GSH: Your Body's Most Powerful Protector.* 3rd ed. Montreal: Kudo.Ca Communications, 2002, 206.

Ikeno, Hiroshi, Takumi Tochio, Hiroshi Tanaka, and Satoru Nakata. "Decrease in Glutathione May Be Involved in Pathogenesis of Acne Vulgaris." *Journal of Cosmetic Dermatology* 10, no. 3 (September 2011): 240–244.

Landry, Ali. Email message to author, January 31, 2020.

"Melanocyte." *Encyclopaedia Britannica.* Accessed February 15, 2020. https://www.britannica.com/science/melanocyte.

Mills, Otto H., Maressa C. Criscito, Todd E. Schlesinger, Robert Verdicchio, and Ernest Szoke. "Addressing Free Radical Oxidation in Acne Vulgaris." *Journal of Clinical and Aesthetic Dermatology* 9, no. 1 (January 2016): 25–30.

Pai, Deanna. "So, It Looks Like Free Radicals Are to Blame for Acne Too." *Glamour*, July 14, 2016. https://www.glamour.com/story/antioxidants-acne-treatment.

Poljšak, Borut, Raja G. Dahmane, and Aleksandar Godić. "Intrinsic Skin Aging: The Role of Oxidative Stress." *Acta Dermatovenerologica Alpina, Pannonica, et Adriatica* 21, no. 2 (January 1, 2012): 33–36.

"Psoriasis." Mayo Clinic. Accessed February 15, 2020. https://www.mayo clinic.org/diseases-conditions/psoriasis/symptoms-causes/syc-20355840.

Seifert, O., J. Holmberg, and B. M. Linnarsson. "Decreased Activity of Neutrophil Glutathione Peroxidase in Chronic Plaque-Type Psoriasis." *Journal of the American Academy of Dermatology* 57, no. 3 (2007): 528–529.

Tyrrell, Rex M., and Mireille Pidoux. "Endogenous Glutathione Protects Human Skin Fibroblasts Against the Cytotoxic Action of Uvb, Uva and Near-Visible Radiations." *Photochemistry and Photobiology* 44, no. 5 (November 1986): 561–564.

"UV Radiation & Your Skin." Skin Cancer Foundation. Accessed February 15, 2020. https://www.skincancer.org/risk-factors/uv-radiation/.

Watanabe, Fumiko, Erika Hashizume, Gertrude P. Chan, and Ayako Kamimura. "Skin-Whitening and Skin-Condition-Improving Effects of Topical Oxidized Glutathione: A Double-Blind and Placebo-Controlled Clinical Trial in Healthy Women." *Clinical, Cosmetic and Investigational Dermatology* 7 (January 1, 2014): 267–274.

Weschawalit, Sinee, Siriwan Thongthip, Phanupong Phutrakool, and Pravit Asawanonda. "Glutathione and Its Antiaging and Antimelanogenic Effects." *Clinical, Cosmetic and Investigational Dermatology* 10 (January 1, 2017): 147–153.

"What Is a Skin Cycle?" Medifine Aesthetics, May 17, 2017. https://www.medifine.co.uk/what-is-a-skin-cycle/.

Zhang, Zheng, Ophir Ortiz, Ritu Goyal, and Joachim Kohn. "Biodegradable Polymers." In *Principles of Tissue Engineering*, 441–473. Edited by R. P. Lanza, Robert Langer, and Joseph P. Vacanti. Burlington, VT: Elsevier Science, 2014.

CHAPTER 8

Ali, Ather, Valentine Yanchou Njike, Veronika Northrup, Alyse B. Sabina, Anna-Leila Williams, Lauren S. Liberti, Adam I. Perlman, Harry Adelson, and David L. Katz. "Intravenous Micronutrient Therapy (Myers' Cocktail) for Fibromyalgia: A Placebo-Controlled Pilot Study." *Journal of Alternative and Complementary Medicine* (New York) 15, no. 3 (March 1, 2009): 247–257.

Gaby, A. "Intravenous Nutrient Therapy: The 'Myers' Cocktail.'" *Alternative Medicine Review: A Journal of Clinical Therapeutic* 7, no. 5 (2002): 389–403.

Hauser, Robert A., Kelly E. Lyons, Terry McClain, Summer Carter, and David Perlmutter. "Randomized, Double-Blind, Pilot Evaluation of Intravenous Glutathione in Parkinson's Disease." *Movement Disorders* 24, no. 7 (May 15, 2009): 979–983.

Millam, D. "The History of Intravenous Therapy." *Journal of Intravenous Nursing: The Official Publication of the Intravenous Nurses Society* 19, no. 1 (1996): 5–14.

Sechi, Gianpietro, Maria G. Deledda, Guido Bua, Wanda M. Satta, Giovanni A. Deiana, Giovanni M. Pes, and Giulio Rosati. "Reduced Intravenous Glutathione in the Treatment of Early Parkinson's Disease." *Progress in Neuropsychopharmacology & Biological Psychiatry* 20, no. 7 (1996): 1159–1170.

ACTION PLAN ONE

"Adaptation of Crops to Increased Carbon Dioxide and Warming." US Department of Agriculture, last updated February 14, 2020. https://www.ars.usda.gov/research/project/?accnNo=430945.

Afshin, Ashkan, Patrick John Sur, Kairsten A. Fay, Leslie Cornaby, Giannina Ferrara, Joseph S. Salama, Erin C Mullany, et al. "Health Effects of Dietary Risks in 195 Countries, 1990–2017: A Systematic Analysis for the Global Burden of Disease Study 2017." *Lancet* 393, no. 10184 (May 11, 2019): 1958–1972.

Heid, Markham. "Is Ghee Healthy? Here's What the Science Says." *Time*, April 22, 2019, https://time.com/5571810/is-ghee-healthy/.

"How Much Protein Do You Need Every Day?" Harvard Health Publishing, last updated June 25, 2019. https://www.health.harvard.edu/blog/how-much-protein-do-you-need-every-day-201506188096.

Hyman, Mark. "Why Vegetable Oils Should Not Be Part of Your Diet." EcoWatch, February 2, 2016. https://www.ecowatch.com/dr-mark-hyman-why-vegetable-oils-should-not-be-part-of-your-diet-1882164589.html.

Jones, Dean P., Ralph J. Coates, Elaine W. Flagg, John W. Eley, Gladys Block, Raymond S. Greenberg, Elaine W. Gunter, and Bethany Jackson. "Glutathione in Foods Listed in the National Cancer Institute's Health Habits and History Food Frequency Questionnaire." *Nutrition and Cancer* 17 no. 1 (1992): 57–75.

Malik, Vasanti. "Is There a Place for Coconut Oil in a Healthy Diet?" Harvard Health Publishing, January 14, 2019. https://www.health.harvard.edu/blog/is-there-a-place-for-coconut-oil-in-a-healthy-diet-2019011415764.

Matta, Murali K., Robbert Zusterzeel, Nageswara R. Pilli, Vikram Patel, Donna A. Volpe, Jeffry Florian, Luke Oh, et al. "Effect of Sunscreen Application Under Maximal Use Conditions on Plasma Concentration of Sunscreen Active Ingredients: A Randomized Clinical Trial." *JAMA* 321, no. 21 (June 4, 2019): 2082–2091.

Milder, Julie, and Manisha Patel. "Modulation of Oxidative Stress and Mitochondrial Function by the Ketogenic Diet." *Epilepsy Research* 100, no. 3 (July 2012): 295–303.

Mozaffarian, Dariush, and Dan Glickman. "Opinion: Our Food Is Killing Too Many of Us." *New York Times*, August 26, 2019. https://www.nytimes.com/2019/08/26/opinion/food-nutrition-health-care.html.

"Protein and the Athlete — How Much Do You Need?" American Dietetic Association, July 17, 2017. https://www.eatright.org/fitness/sports-and-performance/fueling-your-workout/protein-and-the-athlete.

Rose, Karen. "Your Reef Safe Sunscreen Guide — 15 Sunscreens That Are Reef Safe." Hawaii. Accessed February 2, 2020. https://www.hawaii.com/blog/reef-safe-sunscreen/.

"Selenium." National Institutes of Health Office of Dietary Supplements, last updated October 17, 2019. https://ods.od.nih.gov/factsheets/Selenium-HealthProfessional/.

"Sunscreen Drug Products for Over-the-Counter Human Use." *Federal Register*, February 26, 2019. https://www.federalregister.gov/documents/2019/02/26/2019-03019/sunscreen-drug-products-for-over-the-counter-human-use.

ACTION PLAN TWO

"Can Coffee Help Your Liver?" WebMD. Accessed February 15, 2020. https://www.webmd.com/hepatitis/coffee-help-liver#1.

Carbajo, Jose Manuel, and Francisco Maraver. "Sulphurous Mineral Waters: New Applications for Health." *Evidence-Based Complementary and Alternative Medicine* 2017 (2017): 11.

"Chromium." National Institutes of Health Office of Dietary Supplements, last updated July 9, 2019. https://ods.od.nih.gov/factsheets/Chromium-HealthProfessional/.

"Copper." National Institutes of Health Office of Dietary Supplements, last updated July 19, 2019. https://ods.od.nih.gov/factsheets/Copper-Health Professional/.

"FDA Drug Safety Communication: Prescription Acetaminophen Products to Be Limited to 325 mg per Dosage Unit; Boxed Warning Will Highlight Potential for Severe Liver Failure." US Food and Drug Administration, last updated February 7, 2018. https://www.fda.gov/drugs/drug-safety-and-availability/fda-drug-safety-communication-prescription-acetaminophen-products-be-limited-325-mg-dosage-unit.

"Folate." National Institutes of Health Office of Dietary Supplements, last updated July 19, 2019. https://ods.od.nih.gov/factsheets/Folate-Health Professional/.

Ji, Deng-Bo, Jia Ye, Chang-Ling Li, Yu-Hua Wang, Jiong Zhao, and Shao-Qing Cai. "Antiaging Effect of Cordyceps Sinensis Extract." *Phytotherapy Research* 23, no. 1 (January 2009): 116–122.

Kıvrak, Elfide Gizem, Kıymet Kübra Yurt, Arife Ahsen Kaplan, Işınsu Alkan, and Gamze Altun. "Effects of Electromagnetic Fields Exposure on the Antioxidant Defense System." *Journal of Microscopy and Ultrastructure* 5, no. 4 (December 2017): 167–176.

"Lipoic Acid." Linus Pauling Institute, Oregon State University. Accessed February 4, 2020. https://lpi.oregonstate.edu/mic/dietary-factors/lipoic-acid.

Lubkowska, Anna, Barbara Dołęgowska, Zbigniew Szyguła, and Stephane Blanc. "Whole-Body Cryostimulation—Potential Beneficial Treatment for Improving Antioxidant Capacity in Healthy Men—Significance of the Number of Sessions (Whole-Body Cryostimulation on Antioxidant System)." *PLoS ONE* 7, no. 10 (October 15, 2012): e46352.

Michna, Edward, Mei Sheng Duh, Caroline Korves, and June L. Dahl. "Removal of Opioid/Acetaminophen Combination Prescription Pain Medications: Assessing the Evidence for Hepatotoxicity and Consequences of Removal of These Medications." *Pain Medicine* 11, no. 3 (March 2010): 369–378.

Miltonprabu, Selvaraj, Michał Tomczyk, Krystyna Skalicka-Woźniak, Luca Rastrelli, Maria Daglia, Seyed Fazel Nabavi, Seyed Moayed Alavian, and Seyed Mohammad Nabavi. "Hepatoprotective Effect of Quercetin: From Chemistry to Medicine." *Food and Chemical Toxicology* 108, no. Pt B (October 2017): 365–374.

Panahi, Yunes, Parisa Kianpour, Reza Mohtashami, Stephen L. Atkin, Alexandra E. Butler, Ramezan Jafari, Roghayeh Badeli, and Amirhossein Sahebkar. "Efficacy of Artichoke Leaf Extract in Non-alcoholic Fatty Liver Disease: A Pilot Double-Blind Randomized Controlled Trial." *Phytotherapy Research* 32, no. 7 (July 2018): 1382–1387.

"Vitamin B6." National Institutes of Health Office of Dietary Supplements, last updated September 19, 2019. https://ods.od.nih.gov/factsheets/VitaminB6 -HealthProfessional/.

"Vitamin B12." National Institutes of Health Office of Dietary Supplements, last updated July 9, 2019. https://ods.od.nih.gov/factsheets/VitaminB12 -HealthProfessional/.

"Vitamin E." National Institutes of Health Office of Dietary Supplements, last updated July 10, 2019. https://ods.od.nih.gov/factsheets/VitaminE -HealthProfessional/.

"What Is Chelation Therapy?" WebMD. Accessed February 4, 2020. https:// www.webmd.com/balance/guide/what-is-chelation-therapy#2.

Zembron-Lacny, A., M. Slowinska-Lisowska, Z. Szygula, Z. Witkowski, and K. Szyszka. "Modulatory Effect of N-Acetylcysteine on Pro-Antioxidant Status and Haematological Response in Healthy Men." *Journal of Physiology and Biochemistry* 66, no. 1 (March 2010): 15–21.

"Zinc." National Institutes of Health Office of Dietary Supplements, last updated July 10, 2019. https://ods.od.nih.gov/factsheets/Zinc-Health Professional/.

ACTION PLAN THREE

Ashoush, Sherif, Amgad Abou-Gamrah, Hassan Bayoumy, and Noura Othman. "Chromium Picolinate Reduces Insulin Resistance in Polycystic Ovary Syndrome: Randomized Controlled Trial." *Journal of Obstetrics and Gynaecology Research* 42, no. 3 (March 2016): 279–285.

Draelos, Zoe Kececioglu, Jeffrey S. Dover, Murad Alam, Miles Hitchen, and John Bajor. *Cosmeceuticals.* 3rd ed. St. Louis, MO: Elsevier, 2016.

Feiner, J. J., M. A. Mcnurlan, R. E. Ferris, D. C. Mynarcik, and M. C. Gelato. "Chromium Picolinate for Insulin Resistance in Subjects with HIV Disease: A Pilot Study." *Diabetes, Obesity and Metabolism* 10, no. 2 (February 2008): 151–158.

Jeong. Email message to author, November 12, 2019.

Narda, Mridvika, Laurent Peno-Mazzarino, Jean Krutmann, Carles Trullas, and Corinne Grange. "Novel Facial Cream Containing Carnosine Inhibits Formation of Advanced Glycation End-Products in Human Skin." *Skin Pharmacology and Physiology* 31, no. 6 (January 1, 2018): 324–331.

Ryland. Interview with author, November 12, 2019.

"A Scientific Review: The Role of Chromium in Insulin Resistance." *Diabetes Educator* Supplement (2004): 2–14.

Weiss R. A., and M. A. Weiss. "Evaluation of a Novel Anti-Aging Topical Formulation Containing Cycloastragenol, Growth Factors, Peptides, and Antioxidants." *Journal of Drugs in Dermatology* 13, no. 9 (2014): 1135–1139.

AFTERWORD

"Human Endocannabinoid System." UCLA Health. Accessed February 16, 2020. https://www.uclahealth.org/cannabis/human-endocannabinoid-system.

"NADH." WebMD. Accessed February 16, 2020. https://www.webmd.com/vitamins/ai/ingredientmono-1016/nadh.

INDEX